Modelling the Labour Market

INTERNATIONAL STUDIES IN ECONOMIC MODELLING

Series Editor

H. Motamen

Executive Director,
Canadian Imperial Bank of Commerce,
Securities Europe Ltd,
London

Economic Modelling in the OECD Countries
H. Motamen
Input-output Analysis
M. Ciaschini

In preparation

Models of Disequilibrium and Shortage in the Centrally Planned Economies
C. Davis and W. Charemza
Economic Models of Latin America
D.E. Hojman
Economic Models of Trade Unions
P. Garonna

Modelling the Labour Market

Edited by
Michael Beenstock
Lady Davos Professor of Economics,
Hebrew University, Jerusalem

London
CHAPMAN AND HALL
New York

First published in 1988 by
Chapman and Hall Ltd
11 New Fetter Lane, London EC4P 4EE
Published in the USA by
Chapman and Hall
29 West 35th Street, New York, NY 10001

MyCopy version of the original edition 1988

British Library Cataloguing in Publication Data

Modelling the labour market. – (International
studies in economic modelling).
1. Labour supply – Great Britain –
Econometric models
I. Beenstock, Michael II. Series
331.12'0941 HD5765.A6

DOI 10.1007/978-94-009-1203-8

Library of Congress Cataloging in Publication Data

Modelling the labour market/edited by Michael Beenstock.
p. cm. – (International studies in economic modelling)
Includes bibliographies and index.

1. Labor supply – Great Britain – Econometric models.
I. Beenstock, Michael, 1946– . II. Title: Modelling the labor
market. III. Series.
HD5765.A6M63 1988
331.12'0724 – dc19 87-23084
CIP

www.springer.com/mycopy

Contents

	Contributors	vii
	Introduction to the Series	viii
	Preface	x
1	**The labour market: theory and experience**	**1**
	Christopher Bliss	
1.1	The concept of the labour market	1
1.2	The supply and demand for labour	2
1.3	Models of unemployment	7
1.4	Macroeconomics and the labour market	14
1.5	The Phillips curve and the NAIRU	17
1.6	Some features of the data	19
	References	22
2	**Some formal models of the aggregate labour market**	**25**
	Martyn Andrews	
2.1	Introduction	25
2.2	Labour demand	26
2.3	Some formal models	30
	Appendix A	44
	References	46
3	**The disequilibrium approach to modelling the labour market**	**49**
	Stephen Hall and Brian Henry	
3.1	Introduction	49
3.2	The multimarket non-clearing model	52
3.3	The labour market in disequilibrium	54
3.4	Empirical results	60
3.5	Policy implications	65
3.6	Conclusions	68
	References	68
4	**A neoclassical model of the UK labour market**	**71**
	Michael Beenstock and Peter Warburton	

4.1 Introduction 71
4.2 Theory 72
4.3 Econometric results 81
4.4 Simulations 89
Appendix A 100
References 103

5 A new classical model of the labour market 105
Patrick Minford
5.1 The model in outline 106
5.2 The labour market in detail 110
5.3 The demand for labour 120
5.4 Responses to critics 139
5.5 Conclusions 142
Acknowledgement 143
References 143

6 Imperfect competition and the labour market 145
Stephen Nickell
6.1 Introduction 145
6.2 Price and employment behaviour 145
6.3 Wage determination 149
6.4 The labour market in a macromodel 152
6.5 Empirical analysis of the British labour market 160
6.6 Summary and conclusions 166
References 166

7 An international perspective 169
Martyn Andrews
7.1 Introduction 169
7.2 Multi-country studies: some comments 170
7.3 Selected applications of labour market models 173
7.4 Multi-country studies: some examples 184
7.5 Institutional characteristics 188
7.6 Concluding comments 193
Acknowledgements 195
References 195

Index 199

Contributors

Martyn Andrews	Department of Econometrics and Social Statistics, University of Manchester, UK
Michael Beenstock	Hebrew University, Jerusalem
Christopher Bliss	Nuffield College Oxford, UK
Stephen Hall	Economics Division, Bank of England, London, UK
Brian Henry	National Institute of Economic and Social Research, London, UK
Patrick Minford	Department of Economics and Accounting, University of Liverpool, UK
Stephen Nickell	Institute of Economics and Statistics, Oxford, UK
Peter Warburton	Shearson Lehman, London, UK

Introduction to the Series

There has been a growing dependence in the past two decades on modelling as a tool for better understanding of the behaviour of economic systems, and as an aid in policy and decision making. Given the current state of the art globally, the introduction of a series such as this can be seen as a timely development. This series will provide a forum for volumes on both the theoretical and applied aspects of the subject.

International Studies on Economic Modelling is designed to present comprehensive volumes on modelling work in various areas of the economic discipline. In this respect one of the fundamental objectives is to provide a medium for ongoing review of the progression of the field.

There is no doubt that economic modelling will figure prominently in the affairs of government and in the running of the private sector, in efforts to achieve a more rational and efficient handling of economic affairs. By formally structuring an economic system, it is possible to simulate and investigate the effect of changes on the system. This in turn leads to a growing appreciation of the relevance of modelling techniques. Our aim is to provide sufficient space for authors to write authoritative handbooks, giving basic facts with an overview of the current economic models in specific areas and publish a useful series which will be consulted and used as an accessible source of reference.

The question may arise in some readers' minds as to the role of this series *vis-à-vis* other existing publications. At present, no other series possesses the characteristics of *International Studies on Economic Modelling* and as such cannot fill the gap that will be bridged by it, while those journals which focus in this area do not present an exhaustive and up-to-date overview of a particular subject and all the developments in the field. Other journals which may contain economic modelling papers are not sufficiently broad to publish volumes on all aspects of economic modelling which this series is designed to cover.

A variety of topics will be included encompassing areas of both micro and macroeconomics, as well as the methodological aspects of model construction. Naturally, we are open to suggestions from all readers of, and contributors to, the series regarding its approach and content.

Finally, I would like to thank all those who have helped the launch of this series. The encouraging response received from authors who have contributed the forthcoming volumes and from the subscribers to the series has indicated the need for such a publication.

Homa Motamen
London, Dec 1987

Preface

It is generally agreed that the operation of the labour market lies at the heart of macroeconomic activity. Following Keynes' attack on the postulates of the classics in *The General Theory* a number of labour market paradigms have been proposed including the Phillips curve, new classical and union bargaining models. These paradigms usually form the centrepiece of any good text on macroeconomic theory. Our purpose in this volume is not to restate these paradigms but to explore the extent to which they might be empirically modelled. To this end the volume includes a set of econometric models of the UK labour market where each contribution relates very closely to one of the principal paradigms.

The purpose of this collection is threefold. First and foremost we wanted to present an integrated set of case studies in applied econometrics with reference to labour market modelling. In doing so we hope the volume will appeal to third year undergraduates and postgraduate students studying applied econometrics and labour economics. Secondly, the contributions have been carefully selected to illustrate the main paradigms since each contribution is intellectually self-contained. It is arguable that this may be disadvantageous as far as the truth is concerned if eclecticism is preferable. On the other hand it has pedagogic advantages in drawing sharp distinctions between the various approaches.

The third and least important purpose of the collection is policy design. We make no attempt to judge which of the models is the best. Non-nested testing of individual equations is difficult enough; non-nested testing of entire models raises methodological questions that lie well beyond our present terms of reference. All the models purport to explain similar phenomena such as the growth in unemployment and it is left to the reader to judge the appropriateness of the competing models. This is all the more apposite because to a large extent the models are constructed from broadly similar time-series aggregates.

The research projects that lie behind the contributions were not planned in advance. But better co-ordination could most probably not have been achieved had one tried. Stephen Hall and Brian Henry have tried to estimate a 'Keynesian' model of the labour market in which employment is equal to the smaller of labour supply or demand. However, the demand for labour is constrained by expected aggregate demand but is also influenced by factor prices. In a strictly Keynesian model money wages are rigid downwards. Nevertheless, in Hall and

Henry's model money wages are merely assumed to be sticky and to respond to market forces over time.

Whereas Hall and Henry assume that the labour market does not clear in the short run Patrick Minford assumes instantaneous (within a year) market clearing in his model which is cast in a new classical mode. In this case firms are never demand constrained and wages continuously adjust to balance the expected supply and demand for labour. However, wage contracting implies that money wages are fixed during the current period. On the other hand, both models assume that expectations are 'rational' and this confirms the well-known result that 'rational' expectations are not the *sine qua non* of new classical macroeconomics but instantaneous market clearing is the crucial factor.

The Phillips curve has been an essential element of the so-called 'neoclassical synthesis'. Michael Beenstock and Peter Warburton explore how far the Phillips curve may be used as a basis for modelling the labour market. They also lay greater emphasis on the determinants of the working population. Their model implies that wages adjust over time to equilibrate the labour market but in the short run the demand for labour may differ from its supply. In this their efforts have much in common with Hall and Henry. On the other hand, they do not maintain that the demand for labour is constrained by the expected level of effective demand. Instead they suggest that the demand for labour evolves on neoclassical lines.

Stephen Nickell's model distinguishes itself by assuming that the product markets are imperfectly competitive and that wages are set by trade unions rather than by market forces. The former assumption implies that the demand for labour might be influenced by the level of aggregate demand. This sounds Keynesian but it is not. Instead it appeals to the hypothesis that the price elasticity of demand for products is not independent of the level of aggregate demand. The role of unions in Nickell's model has much in common with Minford's model. However, they measure union power in quite different ways. Also Minford assumes expectations are 'rational' while Nickell does not.

Ideally it would have been desirable to round off the volume with a contribution that sought to encompass all the paradigms that are considered. According to the encompassing principle one model encompasses another when it explains not only where the other model was correct but also where it erred. But in our present context there are at least two problems with this principle. The first is practical; encompassing principles are more difficult to implement for systems of equations than for single equations. Secondly, the competing paradigms are conceptually non-nested and so one cannot meaningfully form an eclectic paradigm from non-nested components.

The volume is structured as follows. Chapters 1 and 2 set the scene. Christopher Bliss reviews labour market theory and discusses the degree to which the applied econometricians have drawn on the available class of theory in setting up their models. His objective is not a comprehensive review of labour economics but a selective appraisal of the relationship between theory and practice. He also introduces the UK data that the econometric modellers seek to explain in the main part of the volume.

In Chapter 2 Martyn Andrews presents an overview of the taxonomy of labour market models that applied econometricians typically estimate. He also concludes the volume by reviewing efforts to model the labour market outside the UK. Chapters 3 to 6 consist of the models themselves.

The contents of this volume relate to narrowly drawn terms of reference. By 'labour market model' we refer to a system of equations that purport to explain wages, employment and unemployment on a consistent basis. This rules out inclusion of perhaps very interesting single equation material on for example wages in isolation. Secondly, we focus exclusively on models estimated from time-series data rather than cross-section data. At the present time, however, an econometric model of the labour market constructed from such data does not exist.

Finally, as editor of the volume, it remains for me to thank the various contributors who enthusiastically responded to my suggestion to bring together our disparate efforts in this way.

Michael Beenstock
March 1987

1

The labour market: theory and experience

CHRISTOPHER BLISS

1.1 THE CONCEPT OF THE LABOUR MARKET

So much has the term 'the labour market' become a part of the language of contemporary political debate that its meaning might appear to require no explanation. However, the phrase conceals important and interesting questions. Two issues in particular stand out.

First it is important to note what is meant by a 'market'. What is intended is not an identifiable institution at a specific location, such as the Billingsgate Fish Market, but something more abstract. The market is a loose agglomeration of numerous bargains, not strictly connected yet not wholly unrelated either.

Thus the concept of a 'market' for labour or some other good makes use of a device which we shall refer to as *the aggregation of transactions*. It assumes that all the transactions involving a particular good are based on the same information and are coordinated by the same forces, the pressures of supply and demand, and that common prices get established.

The second major issue is also a question of aggregation, in this case *the aggregation of different markets*. The British labour market is clearly a large collection of different markets concerned *inter alia* with the labour of gardeners in the South-East, plasterers in Scotland, actors and doctors. All these different labour markets are interrelated in a manner which cannot be ignored but which is exceedingly complicated to describe.

Given these difficulties, should we even consider labour markets defined in a highly aggregated manner? This type of issue is fairly familiar in economics, and is not one to which one could hope to provide a definite and simple solution. However, not all the arguments tell against aggregation. There are two other ways of proceeding. One would be a fully disaggregated general equilibrium analysis in which the details of individual bargains together with their multifarious interconnections would all be taken into account. This sounds ideal, but as a research strategy it is of course impossible.

Given the impracticality and doubtful worth of a complete general equilibrium approach, the only serious alternative to aggregation is partial equilibrium analysis. This has its uses but there are also questions for which the partial equilibrium approach is not helpful. In particular, partial equilibrium analysis cannot throw very much light on the determination of wages in general, as opposed to the wage rate in a particular occupation. A general equilibrium view of the labour market is then inescapable.

Sometimes bargaining is between employers and unions. Bargaining of this type is rather common, particularly in the British labour market in which compared with other industrial countries a large proportion of employees belong to a trade union. It is essentially arbitrary whether or not we classify the arena of large agent negotiation as a market, but there is something to be said for doing so.

In discussing the meaning of a market we mentioned supply and demand as playing a role in defining the market. Supply and demand curves assume competitive behaviour by buyers and sellers. However, even in a market in which one side, or both sides, of transactions are dominated by one, or a few, large agents may be conceived of as having a supply and demand background. This is to say that we may always ask how agents would transact if they ignored their influence on the market price. Being large agents they will not do that, yet it may still be important to know the supply and demand curves for they contain a great deal of information.

1.2 THE SUPPLY AND DEMAND FOR LABOUR

We start with the demand for labour. This, like the supply of labour, may be treated by a fairly simple static model, or by means of a more dynamic approach. In either case the demand for labour is derived from profit maximization. According to which inputs are taken as fixed, say the capital stock or the number of firms, we have short-run profit maximization or various degrees of long-run profit maximization.

For a general treatment it is useful to work with the net revenue function of the firm. Net revenue, denoted $R(E)$, is defined as the maximum excess of sales value over costs of inputs, including any inputs that may be varied, that the firm can achieve when it employs E workers. The net revenue may depend only on prices, as would be the case with a perfectly competitive small firm, or may include a parameter representing the level of demand. We assume however that the firm is a price-taker in the labour market so that it can hire as many workers as it pleases at a cost of w for each worker. The profit maximizing employment level is then the solution to:

$$\max_{E} R(E) - wE. \tag{1.1}$$

It is easily shown that an increase in the wage rate can only discourage employment and that this is true independently of the shape of the $R(E)$ function which may include non-concave sections. Consider two wage rates, w_1 and w_2, with $w_2 > w_1$, and corresponding profit maximizing employment levels, E_1 and E_2. As these employment levels maximize when the corresponding wage rules, we must have:

$$R(E_1) - w_1E_1 > R(E_2) - w_1E_2, \tag{1.2a}$$

and

$$R(E_2) - w_2E_2 > R(E_1) - w_2E_1. \tag{1.2b}$$

From which it follows that:

$$0 \geqslant (E_1 - E_2)(w_1 - w_2) \tag{1.3}$$

and, as the term in the second bracket is negative, by assumption, the term in the first bracket must be non-negative. Hence an increase in w cannot have encouraged employment, as required.

Dynamic extensions of the above model include labour hoarding, employer search and adaptive adjustment towards the optimum. We shall touch on employer search along with employee search in Section 1.3 below. For adaptive adjustment in the form of an error correction mechanism see Beenstock and Warburton (1982).

Suppose that the revenue function contains a parameter u which varies over time and which may be taken to represent the level of demand for, or the price of, the output produced by the firm. To each value of u there corresponds a level of employment which would be the solution to expression 1.1 if u were always constant at that level. If adjustment of the labour input were costless, the firm could simply adjust its employment of labour so as to continuously maximize $R(E, u) - wE$. However, in reality there are always costs to adjusting the level of the labour input[1] and these costs are distinct from the costs of employing labour which are included in w. These costs include hiring costs, including training, and firing costs, including redundancy payments. When the firm takes into account that some changes in u are transitory in nature, it will be optimal for it to adjust only partially to a change in u, so that employment will be more stable than is demand as represented by u.

Notice the role played by forward-looking expectations. The demand for labour depends as importantly on the expectation of future demand as on present demand. Evidence of slow adjustment of the labour input to shocks is provided by Beenstock and Warburton (1982) and Deaton (1982).

We turn now to the supply of labour. In the model that commonly underlies the analysis of the labour supply decision the household chooses how much labour to supply taking into account, on the one hand, the opportunity cost of work in terms of leisure foregone and, on the other hand, the benefit in terms of extra

[1] The same point has been argued for the adjustment of the capital input, but we shall not pursue that issue here.

consumption that work for a wage makes available. There is no money illusion, so that it is the real value of the wage that matters.

We shall also postpone for later discussion the point that the wage is fixed in terms of money and that the implied real wage may be uncertain; also the point that the opportunity cost of working may not be so much leisure foregone, but rather the loss of the opportunity to search for a higher wage. We are thus left with the consideration of the static and full information aspects of the decision to work or not to work. However, this simple case is not devoid of interest. It allows us to consider in a simple theoretical model, for example, two important questions. What is the effect of higher unemployment benefits on the supply of labour and hence on unemployment? And, how is the supply of labour by married women related to the wage level?

We first consider a single worker household. Later another worker, the wife, will be added. The utility of the household depends on the amount of leisure enjoyed, L, and on consumption, C, as:

$$U(C, L). \tag{1.4}$$

If the household opts for unemployment[2] it will receive a payment B, the benefit level, and its utility will be $U(B, L_0)$, where L_0 is the maximum leisure possible, say 168 hours per week. If the household elects to be employed, its utility will be the solution to:

$$\max_L U[(L_0 - L)w, L], \tag{1.5}$$

where w is the wage rate per unit of leisure surrendered to work, which we may denote by $V[w]$. If work involves a fixed value of L, say 40 hours per week, then $V[w]$ is simply $U[40w, 128]$. The household chooses employment when:

$$V[w] \geqslant U(B, 168). \tag{1.6}$$

The maximum utility which the household may enjoy, $V[w]$ or $U(B, 168)$ whichever is the greater, may be denoted $V[w, B]$. This function, and particularly its partial derivatives, which are the quantities of labour supplied,[3] plays an important role in explaining both the amount of work done by those who work, and the numbers electing not to work.

Consider a household in employment and a small variation in the wage rate, w. If the variation is an increase, the household will clearly be better off, as it

[2] In Britain, as in other countries, eligibility for unemployment benefit depends upon the worker being willing to take a job. However, this rule has not always been rigorously enforced. Layard (1986, pp. 50–3) argues that enforcement became slacker over time with a notable relaxation in the 1970s.

[3] The partial derivative of an indirect utility function is equal to the product of the marginal utility of income and the quantity demanded or supplied: see Varian (1978, p. 116). However, the marginal utility of the wage is undefined at the point where the household is indifferent between working and not working, and the left-hand derivative is zero, as falls in the wage below the level necessary to induce the household to work have no effect on utility. Here both left- and right-hand derivatives are defined and their inequality corresponds to an ambiguity in labour supply at the point at which the household is indifferent between working and claiming benefit.

may work the same amount and enjoy higher consumption. It may choose to work more as a result of the higher wage, but there is no necessity for this. It might take out the benefit of a higher wage partly in increased leisure, the income effect. As is well known, the effect of a higher wage on the amount of labour provided by a household which can regulate its hours of work is ambiguous.

Note, however, that the ambiguity just mentioned applies only to the working household, for it is only the working household that benefits from an increase in the wage rate, hence only the working household that enjoys an income effect from an increase in the wage which may be taken out in increased leisure. The effect of an increase in the wage rate on the participation decision, the decision to work or not to work, is unambiguous. An increase in the wage rate will lead to $V[w]$ exceeding $U(B, 168)$ for some households for which previously it was the other way round. As $U(B, 168)$ is constant, and only $V[w]$ is affected by the wage rate, an increase in the wage rate can only increase employment. As participation decisions are of much greater importance for the total quantity of labour supplied than are variations in the amount of work supplied by an employed worker, which in many cases is imposed by the employer, this is an important conclusion.

An analogous argument to the one that has just been rehearsed for an increase in the wage rate applies to a decrease in the benefit level, B. This decreases $U(B, 168)$ and has no influence on $V[w]$, so it can only tip the balance in favour of participation. We have therefore justified the roles assigned in the analysis of labour supply to the wage rate, possibly conceived as total earnings from a standard week's work, and the benefit level.

Notice, however, that we have not arrived at the replacement ratio, which is the statistic most commonly used to represent the effect of benefit on the supply of labour. The replacement ratio is the amount of income which a household would receive as benefit divided by the net income after tax that it would receive in work. The supply of labour is the integral of the households for which utility in work exceeds utility when unemployed. This is related of course to both earnings in work and replacement income from benefit, but not particularly to the ratio of the two.

To exactly justify a concentration on the replacement ratio we would need to assume the utility function to be linear in the logarithms of consumption and leisure, as $U(C, L) = a \log C + (1 - a) \log L$, because then increasing C by a given proportion while holding L constant adds the same amount to the utility of work as it adds to the utility of unemployment, and any equality between the two is retained. Clearly the use of the replacement ratio implies a linear (in logarithms) approximation to the utility function which is unlikely to be exact[4] but which

[4] If the worker can choose L in work rather than having a fixed working week imposed, the condition as we have written it seems to require modification. However, with a linear in logarithms utility function, the worker chooses to work a week of $(1 - a)L_0$ hours whatever the wage, so that the linear in logarithms form is again sufficient for the replacement ratio to measure the attractiveness of unemployment against work.

is no worse than similar linear approximations that regularly figure in econometric specifications.

We next turn to the supply of labour by married women. To treat the couple let the household have a utility function when in work of the form $U(C, L_f, L_m)$, where L_s (s = f, m) is the amount of leisure available to respectively the female and the male party. It seems correct to treat the utility of female and male leisure (which will include work about the house and garden) as separate factors, and not consider only the sum of the two.[5]

We consider the case in which the husband is in work and the wife decides whether to work or not. We allow for different wage rates for the work done by husband and wife, denoted w_f and w_m. Let the husband be working a 40 hour week. Then the wife's optimal number of hours of work, $168 - L_f$, maximizes:

$$U[40w_m + (L_0 - L_f)w, L_f, 128]. \tag{1.7}$$

Formally this generates a labour supply function just like the supply function for male labour; however, there is an important difference. The wife, the second worker, does not enjoy the possibility of opting for unemployment and receiving in consequence a considerable benefit payment. This provides an added impetus to work where the wife is concerned, but against it must be set the fact that the family is better off by virtue of the husband's earnings than would be a single earner family considering whether any work should be undertaken. As both types of leisure are normal goods, the higher base level of income generates a demand for leisure, in this case household, which may lead the wife to elect not to work even if she could. Add to this the consideration that the wife's potential wage rate is likely to be lower than her husband's, on account of differences in qualifications and disguised[6] sexual discrimination.

It is not surprising then that the participation rate for married women is lower than that for other categories, although it has risen greatly over time. Also, it is only to be expected that part-time work will prove attractive to second[7] workers as within the family the substitution of income for leisure has already been taken quite far.

The increased participation of married women in the labour force has been one of the notable features of the British labour market in the post-war years. This is the result of many factors and there is no reason to believe that the

[5] Greenhalgh (1980) treats the optimization of female labour supply taking male labour supply as given. It is no surprise that this gives good results in practice.

[6] In Britain certain kinds of overt sexual discrimination in employment are illegal, but there is much scope for paying women less by employing them in different jobs, or taking advantage of their greater tendency to work part-time to discriminate against them.

[7] Any insulting overtones of this term are certainly not intended. Of course women's work is of the greatest importance whether inside or outside the home and not secondary in any value sense. However, if the husband will certainly work, as is often the case, the nature of the wife's decision whether to work outside the home or not is fundamentally affected, as has been explained, and the use of the term second is to mark this analytical difference.

increase in the level of real wages earned by women has been the most important one. The increase in female wage levels has been accompanied after all by a parallel increase in male earnings, and the effect of the latter change on female participation is ambiguous. Also other developments have accompanied and certainly influenced the change in female participation, among them a change in the self-perception of women and a decline in their willingness to be only home-makers. The increased availability of labour-saving devices in the home and of easily prepared food, and finally a decrease in child-bearing have plainly worked in the direction of encouraging women to work. Finally, male unemployment has itself proved a powerful incentive to some married women to go out to work. Indeed some have seen the relative success of women in obtaining work in recent years as a reflection of the greater flexibility of the terms, concerning both wages and hours, on which they have been willing to take employment.

1.3 MODELS OF UNEMPLOYMENT

In this section we briefly review various models of unemployment. This is a large field which by itself could easily occupy a chapter, hence our treatment is necessarily extremely cursory. However, it is hoped that a quick review will be useful. For more thorough surveys the reader is referred to Azariadis (1981), Knight (1987), Lippman and McCall (1976), Oswald (1982) and Stiglitz (1984). We consider the theories under four headings: search theories, efficiency wage theories, implicit contract models and trade union models, with no suggestion that these categories exhaust the field, which they do not.

However, before embarking on the discussion of these various models we may note that we have already encountered a model of unemployment in our discussion of labour supply above. There we saw that a household might sometimes find it optimal not to supply labour but to rely instead on replacement income. Replacement income was taken to be unemployment benefit,[8] but it could also have been whatever resources the household could command on its own or with help from relatives, etc. When a household elects not to take employment we seem to have an example of voluntary unemployment.

What is meant by voluntary unemployment? The term is usually taken to imply an unwillingness to accept a job offer.[9] But should this be any offer, or must the

[8] In Britain unemployment benefit is only payable for the first year of unemployment, after which the household must rely on supplementary benefit.

[9] Keynes (1936) used the term somewhat differently. His idea was that a worker strictly preferring work at current wage levels to unemployment would prefer work even at a slightly lower wage, and would therefore support the scheme of general wage cuts which would put him into employment. This account assumes that relativities between wages in different occupations play an important role in the decision to supply labour and that, as workers generally suppose that wages in other occupations will not fall, they oppose a cut in wages in their own occupation. However, the unemployed worker in Keynes' theory is involuntarily unemployed on our definition, as he or she would accept a job offer at the current wage.

offer be at the appropriate rate for the worker's skill, location, etc., and if so what defines what that rate should be? If the offer only counts if it is at the right level, this may not provide a very useful definition if the level concerned is not clearly defined. Consider a worker who has just become unemployed and immediately receives an offer of work, but a lower wage than the one he or she had been receiving previously, which the worker decides to decline. In some sense this worker is voluntarily unemployed, in that he or she could be in work and has elected not to be. However, the worker is not in a state of long-run equilibrium. There would be no harm in classifying as an instance of voluntary unemployment a worker who could be in work at any wage, except that 'voluntary' carries overtones of culpability which would then be quite misplaced.[10]

In discussing search models we must distinguish between employee search and employer search. The former has received most attention, but agents on either side of the labour market may be conceived as facing a distribution of prices or qualities which they may search at a cost. When worker search is involved we suppose that employers post offers of employment, and we can define the wage so that it measures the net benefit of working for a particular employer, in which case it may cost time and effort to discover the true wage. Employers must have some motive for posting particular wage offers, but as the model will produce the result that vacancies for which higher wages are posted are filled more quickly, a motive is readily available.

An essential assumption of the worker search models is that employment and searching offers are alternative and mutually incompatible activities. In particular, once a worker accepts a job he or she is precluded from continued search. The evidence shows (see Layard (1986, p. 22)) that only a small minority of the unemployed have voluntarily quit previous employment, but among those who have been dismissed, or were never employed, may be included some who are searching in the sense that they have passed up jobs that were available to search for a better one. Hence search models may apply best to explaining how quickly the randomly unemployed become re-employed. Rather than explaining entry to the unemployed state, they are perhaps best used to explain how quickly workers leave that state.

Let the worker start with the belief that the wage offers that he or she samples are drawn randomly from a distribution with cumulative density function $F(w)$.[11] Let his or her utility from being employed depend positively on the wage offer that he or she eventually accepts, w^*, and negatively on the number of offers checked, N, as $U(w^*, N)$. It is assumed that the examination of an offer takes a certain, fixed, amount of time so that the more offers examined the longer the worker will remain unemployed.

We consider the search problem in its sequential form as this seems the most

[10] In fact the idea that obtaining well-paid employment requires search and effort can lead to different moral overtones. The unemployed will include on average more ambitious go-getting workers than those in employment, the resting point of the lethargic and risk averse.

[11] $F(w)$ is the probability that an offer will be at least as good as w.

accurate formulation. The worker does not commit himself or herself in advance to search a specific number of offers but decides at each stage whether to continue searching in the light of his or her experience. We assume that offers searched remain available to be accepted at a later stage so that search can only improve the best offer discovered.[12] The optimum solution to this problem takes the form of a policy rule which determines for each pair of values of N, offers searched, and w^+, the maximum wage so far sampled, whether search will be continued or whether it will terminate in the acceptance of the best offer w^+. This type of rule is equivalent to a minimum best wage offer sufficient to terminate search at stage N, and this minimum acceptable wage is called the reservation wage. In general the reservation wage varies with N.

If $U(w^*, N)$ is additive in w^* and N as:

$$U(w^*, N) = U(w^*) - C(N), \tag{1.8}$$

where $C(N)$ is an increasing function of N such that $C(N+1) - C(N)$ increases with N, we can show that the reservation wage must decrease with the number of trials. To see this note that wage w_0 is accepted when N_0 offers have been sampled if:

$$E\{U(w) - C(N)\} \leqslant U(w_0) - C(N_0), \tag{1.9}$$

where the expectation E is formed subject to the operation of the optimal stopping rule from $N_0 + 1$, and subject to the availability of an offer of w_0. If w_0 were to be the best offer at $N_0 + 1$ we could examine a similar inequality to 1.9 to see whether it should then be accepted. As N would be one trial larger, both the left-hand side and the right-hand side would be smaller, but the left-hand side would be more affected, as larger values of N are involved and for larger values of N, $C(N)$ is increased more by an increase of 1 in the value of N. It is true that a different stopping rule will now apply. However, the inequality would be preserved if costs for all N were increased by $C(N_0 + 1) - C(N_0)$, which is the increase on the right-hand side, and the increase in costs for the left-hand side will be worse than that, regardless of the stopping rule. Hence inequality 1.9 implies:

$$E\{U(w) - C(N+1)\} < U(w_0) - C(N_0 + 1), \tag{1.10}$$

and we have shown that the reservation wage must decrease with the number of trials. As the reservation wage decreases with the number of trials and the best wage sampled can only increase, the probability that an offer will be accepted will go to 1 as N goes to infinity provided that there was initially a positive probability that some wage offer would be accepted.

We now consider the consequences of a shift in $F(w)$, say a shift such that $F(w)$ is increased for each value of w, which corresponds to an improvement in

[12] This is unrealistic, but so is any simple assumption, and this one has the advantage of making for a tractable model.

the general level of wages. We distinguish between two cases. In the first workers know that there has been a shift in $F(w)$ and they know the new distribution. In the second case workers take time to realize that the distribution of w has altered.

The effect on unemployment of a known shift in $F(w)$ is ambiguous. To see this suppose that $F(w)$ shifts so that $F'(w)$, the new distribution, takes the form:

$$F'(w) = F(w \cdot c), \tag{1.11}$$

where $c < 1$ is a constant. The cumulative probability of larger values of w are the same as what used to be the cumulative probability of smaller values. Suppose also that:

$$U(w) = \log(w). \tag{1.12}$$

The effect of the change in $F(w)$ in this case is that a higher wage return may be obtained by applying the same stopping rule with the reservation wage increased by $1/c$. This will leave unaltered the expectation of $C(N)$ and will proportionally increase the expectation of the eventual wage by $1/c$, which will add $-\log(c)$[13] to expected utility. We may now confirm that the optimality conditions for the choice of the reservation wage, expressions like 1.9 are unaffected. Hence search will take place exactly as with the old distribution, except that the reservation wage at each N will be larger, and the duration of unemployment will be unaffected. Of course this is only one case, but as one may so readily construct an example in which search is essentially unaffected, it will be clear that the affect on the average length of search could in general go either way.

It is quite different when workers do not realize that the distribution $F(w)$ has shifted. In that case they come to the market with their old reservation wages, and as the distribution is now different, let it be more favourable, they are more likely to make an early encounter with an offer as good as the reservation level. Hence an increase in demand for labour draws out more supply by reducing the numbers searching. Similarly a decrease in demand would swell the population of searchers with now unrealistic reservation wage levels. Eventually workers will learn that the distribution of w has shifted and the effect on search is ambiguous as we have explained.

As employer search is closely parallel to employee search, we need not develop it at length. The employer searches by placing advertisements in different places, interviewing candidates, etc. The workers are a population willing to accept different wage levels, but in this case the wage should be corrected for worker quality. Exactly similar results apply to the solution. The employer will choose a reservation offer, a maximum wage that he is willing to pay, which may vary through time. Again the consequence of an improvement, from the point of view of the employer, in the distribution of acceptable wages depends on whether the employer knows that there has been a shift.

Search models are a useful way of looking at the short- or medium-term

[13] Note that as $c < 1$, $\log c$ is negative.

adjustment of individual workers to unemployment, and they are helpful in the analysis of the consequences for the whole economy of unanticipated shifts in the demand and supply of labour. They are less useful in explaining long-term unemployment and persistent shifts in levels of unemployment.

Efficiency wage models were originally based on a relationship between the wage level and productivity which was supposed to be due to the effects of improved nutrition in improving the productivity of labour.[14] A relationship of that type holds out the possibility of explaining a long-term downward rigidity of wages and hence perhaps of explaining long-term unemployment. The argument is that a wage cut which more than proportionately reduced the productivity of labour would not be undertaken by a profit maximizing employer, not even in the presence of a reserve army of unemployed willing to work for a lower wage.

To transfer these arguments to advanced countries and, by comparison with those of developing countries, rich workers, requires a different basis for the wage productivity relation from nutrition, which in advanced countries is typically more than adequate. It is easier to make the argument work for one firm than to make it plausible generally. Thus an individual firm might reject the idea of offering the unemployed a lower wage than was paid to those in employment on the ground that it would attract only disgruntled workers who felt that their unhappy position had been exploited, and that the productivity of such unhappy workers would be compromised. However, this type of argument (no firm cuts wages because no firm cuts wages) carries less conviction than one which can argue from a unique equilibrium outcome.

More compelling is the argument that labour market clearing with full employment will not be an equilibrium if it results in excessive labour turnover and slacking on the job.[15] However, behaviour like this may impose high costs on the employer, lowering the net productivity of labour. Hence it is in the interest of any employer to offer a higher wage and to thereby ensure 'loyalty' and 'discipline' (that is, lower turnover and less slacking) in his or her workforce. It follows that an excess supply of labour can be a long-run equilibrium where market clearing cannot be.

The argument just rehearsed invokes imperfect information, in that slacking is assumed not to be costlessly observable, but does not depend on differences between workers being unobservable. It is different therefore from the adverse selection model, of for example Weiss (1980), which is based on the idea that the quality of workers recruited by the firm responds positively to an increase in the wage rate offered.[16] The idea is that workers know more about themselves than an employer can discover, so that those who will work for a low wage may have good reason for being so willing. The adverse selection model requires that somewhere, in a subset of firms or in self-employment, the worker's productivity

14 See Bliss and Stern (1978).

15 See, for example, Shapiro and Stiglitz (1984).

16 The analogy is to adverse selection in medical insurance in which insurance attracts applicants less healthy than the general population.

be observable, for otherwise there is no reason why a low wage should attract anyone.

Efficiency wage models have an advantage over search models in that they can perhaps explain long-term unemployment. However, they go to the opposite extreme in that the unemployment that they explain is a more or less permanent feature of the labour market. They do not help to explain the kind of unemployment, trending upward over periods much longer than a cycle, which has characterized European labour markets since the 1970s.

Implicit contract models were first advanced to explain the apparent paradox that while the wage rate fluctuates little over the trade cycle, employment fluctuates a good deal. This was explained in terms of a long-term relationship between employer and worker in which it was optimal to have the worker bear employment risk rather than being committed to work for a highly variable wage. This argument presumed an opportunity cost of labour close to the average wage level and could not therefore explain involuntary unemployment which leaves the worker much worse off than he would be in employment.[17]

More interesting are implicit contract models which incorporate asymmetric information. The model of Grossman and Hart (1980) is important in that it explains involuntary unemployment in the sense that a worker would prefer to be in work at the wage ruling, and yet the model shows that wage cutting will not be rational or efficient. It is no criticism of this model that its assumptions are somewhat far-fetched, since they appear to define more or less minimal assumptions to obtain the conclusion, and thus define what that conclusion requires.

In the Grossman–Hart model the workers are risk neutral and the employer risk averse.[18] Hence if the state of the world which causes fluctuations in the firm's profitability were to be observable, the employer's receipts would be stabilized and workers' earning would fluctuate. However, when the state of the world is observable only by the employer an incentive for misrepresentation results. The employer will tend to claim that the state of the world is worse than it really is, so as to benefit from the insurance provided by the workers, who are meant to take a cut in wages when business is bad. In the Grossman–Hart solution the firm lays off workers whenever it claims that business is bad, hence proving that it really is the case that its profitability has fallen. The extent to which it lays off workers is regulated so that true revelation of the state of the world is incentive compatible. There is a cost to this arrangement, in that sometimes the productivity of laid-off workers will exceed their marginal disutility of work, but the benefit of some insurance for the employer may outweigh this cost.

The interesting point about this model is that the unemployment that it produces

[17] Another way of explaining this point is to note that stabilizing the wage rate does not in general stabilize the worker's income. When the arrangements for replacing income in unemployment are generous enough to do so, so that is optimal for some workers to be wholly laid off, then the resulting unemployment is not involuntary. See Knight (1987, p. 160).

[18] This is not required but some employer risk aversion plays an essential role in the argument.

is really involuntary, at least in the sense that displaced workers would be better off if employed at the same wage rate, or even a lower rate, than that received by those who stay in work. In another sense the employment is voluntary as constituting part of a long-term agreement. However, it is a standard feature of the implicit contract models that the contracts are meant to be 'implicit'. They may not be reflected in concrete agreements that one could discover in written form.

It is easier to believe that implicit contract models explain occasional examples of, particularly voluntary, unemployment than to be convinced that they explain the great mass of unemployment. One reason why this must be so is that implicit contract models are designed to explain the unemployment that arises over the course of the trade cycle, and that unemployment, as has already been remarked, has been relatively less important in recent years, at least in Europe, than a regularly increasing level of unemployment which survives even upswings in the cycle.

The idea that trade unions are responsible for some, or even most, unemployment has a long history, but only in relatively recent times have explicit models of the bargaining activity of trade unions, and of a tendency for this to generate unemployment, been developed. Macdonald and Solow (1981) provides a review of some earlier work and a good modern discussion of what is required of a union theory which will explain unemployment. See also Oswald (1982).

Trade union models provide an example of a notion with more general applicability, namely of workers as 'insiders'. The idea is that wage bargaining typically takes place between employers and those employed, or their representatives, and that for this reason the outsiders, the unemployed, have a relatively slight influence on the outcome. Trade union leaders, it is argued, are more likely to be voted into office if they obtain high wage settlements for their members, even at a cost to employment, than if they obtain more employment but at a lower wage level. In the standard union bargaining model, the objective of bargaining is represented as a function of both employment and the wage rate.

In the most simple version of these models the union negotiates a wage rate and the employer is allowed to choose the level of employment. This type of bargain is inefficient, because the employer is indifferent between employing or not employing the marginal worker, while the union would strictly prefer him to be employed. For this reason a small cut in the wage rate combined with an increase in employment could make both parties to the bargain better off. Employing a partial equilibrium model, Macdonald and Solow (1980) show that a fall in the demand for labour may lead a monopolistic union to negotiate a higher wage.

This type of model explains the involuntary unemployment of the 'outsiders' as a long-run phenomenum, and also shows that unions might make employment fluctuate over the cycle. It can explain an upward trend in unemployment only if unions become more powerful, or the demand for labour more inelastic, over time, neither of which seems particularly plausible.

1.4 MACROECONOMICS AND THE LABOUR MARKET

Models based on the labour market seem to be at variance with what might be called the macroeconomic explanation of unemployment. Keynesians for example attribute unemployment to a lack of effective demand, which takes the focus of interest away from the labour market. Monetarists, in the form of Milton Friedman and his followers, have typically attributed unemployment to disturbances arising in the monetary sector of the economy.[19] We seem to have here an entirely different type of theory of unemployment from any we have so far considered.

In fact the difference between the macroeconomic approach to unemployment and the labour market approach is more apparent than real. This is because all macroeconomic models are founded in and depend upon a theory of the labour market and how it functions.

Consider the Keynesian approach to unemployment. Although Keynes himself advanced a subtle argument according to which the free adjustment of wages might not lead to a convergence to equilibrium, he failed to spell out the details of this result, and few of his followers, and probably not even Keynes himself, have seriously relied on it. The Keynesian explanation of unemployment has rather centred on nominal wage stickiness.[20] It is the stickiness of money wage rates that enables the labour market to reach a short-period equilibrium despite the fact that an insufficiency of effective demand may have lead to involuntary unemployment.

It appears then that the Keynesian explanation of unemployment is very similar to the old classical idea that unemployment is caused by a too-high wage rate. Partly this is correct but there are important differences. First, the Keynesian approach stresses that the labour market is only one component of an interdependent economy, all the parts of which interact, so that it is wrong to think of the wage rate as being given, and the equilibrium level of employment then being determined in the labour market. Secondly, the Keynesian model stresses the importance of nominal wage bargaining, with the implication that the determination of the real wage is mediated through a process of formation of the price level, the outcome of which is not always accurately perceived by the agents concerned.

This inevitably lead to the criticism, advanced early in post-Keynesian debate, that Keynes' model depended on 'money illusion'. The point was that in the most popular version of the Keynesian model, the Hicks IS/LM model, an expansion in aggregate demand lead producers to produce more output, and hence employ more labour, because it resulted in an inflation of nominal good prices. The

[19] Although persistent unemployment is attributed by Friedman to trade unions and government regulation of wages.

[20] Wage rigidity is the term usually used, but it is not satisfactory, as it seems to imply an absolute immovability that no-one ever intended. We prefer wage stickiness, implying that the wage rate may adjust, but that it does so insufficiently rapidly to continuously clear the market.

result was a decline in the real purchasing power of the given nominal wage which made it profitable for producers to employ more labour. Yet workers were supposed to be unwilling to cut the nominal wage to achieve the same result. Hence the charge of money illusion. This analysis assumed that producers are not rationed in the amount that they can sell so that it needs a change in prices for more output to be produced, a specification later rejected by many Keynesian writers, notably Malinvaud (1980).

Most contemporary Keynesians argue that wage stickiness is a fact, and Solow (1980) has firmly repudiated the suggestion that economists have no right to assume wage stickiness until they have a full explanation of why it occurs. Some writers have invoked the lags in adjustment that are caused by fixed contract lengths to argue that shocks disrupt equilibrium, even though prices are eventually perfectly flexible. For an example of this approach, see Fisher (1977). The consequence of attributing price stickiness to adjustment lags is that Keynesian problems exist but are rather fleeting.

One of the fundamental tenets of Friedman's monetarism[21] is the absence of money illusion. The Keynesian specification of labour supply is rejected and anticipated inflation is held to be neutral in its consequences. It is only unexpected changes in the rate of price inflation that matter for anything, including the supply and demand for labour. Thus Friedman would maintain that the Keynesian method of generating more output and employment by an inflation of output prices[22] can only succeed in the short run, only that is until workers take into account the new prices in deciding how much labour to supply. One consequence of this approach, as we shall see in the next section, was a radical reformulation of the Phillips curve.

It is however a marked feature of Friedman's analysis that monetary shocks can have a real influence on employment in the short-run. Why is this? The most detailed answer was provided by Phelps (1970). The idea is that wage bargains are made in terms of money, and the real value of a money wage depends on the rate of price inflation that intervenes between the closing of the bargain, the work being done, and the worker spending the money. Hence the supply of labour will depend on expectations of inflation, the higher expected inflation the less effective in drawing forth labour is a given nominal wage rate.

A monetary shock in the form of an unexpected increase in the money supply is supposed first to increase demand for output and hence for labour. As producer sales rationing is excluded from consideration, this must imply an eventual increase in prices. Producers are supposed to foresee these price increases better than workers, perhaps because they are closer to pricing decisions, hence their willingness to hire more labour and plan for more production. Workers on the other hand are forecasting inflation using something like adaptive expectations, in which they correct their forecasts in the light of previous errors. So workers

[21] See Friedman (1968) and (1969).

[22] As in the IS/LM model.

read the increase in nominal wages which results from employers' increased demand for labour as an increase in the real wage and for this reason are willing to supply more labour. Later the rate of inflation turns out to be higher than workers had expected, they are disappointed in the real value of their earnings and they adjust their inflation expectations upwards.

In this framework it is only the unexpected component of inflation that draws forth additional labour supply. Another way of putting the same point is to say that it is an acceleration in the rate of inflation that has real effects, the level as such is eventually unimportant. Of course monetarists recognized that rapid rates of inflation might be costly in that they taxed the holding of non-interest-bearing cash holdings, but this was seen as a secondary effect, and one that would be overcome if current account holdings were to pay interest.

Notice that the short-run labour supply response depends on an asymmetry between the price expectations of the buyers and the sellers of labour. If workers and employers both expect a lower rate of inflation than will in fact result, there is no effect on either the demand or the supply of labour, but if labour alone expects a low rate of inflation we get a favourable supply response.

This fundamental asymmetry between information on the two sides of the labour market appears again in the Lucas–Sargent supply function[23] which is at the heart of the new classical macroeconomic model. This incorporates the same assumption that future price increases, or rather shocks to future prices, are better anticipated by producers than by workers, with the consequence that short-run disturbances to prices, always seen as of monetary origin, call forth temporary changes in aggregate supply.

The important difference from the Friedman–Phelps analysis lies in the treatment of expectations. In Friedman these were adaptive, so that a period of learning and adjustment of expectations intervened between the arrival of a shock and its eventual absorption into expectations and consequent neutralization. In the new classical model rational expectations are assumed, so that expectations are supposed to already include every systematic feature of the model. This implies for example that a monetary change will not be unexpected if it is a predictable response of the authorities to the state of the economy, and from this follows the famous conclusion that monetary intervention to influence output and employment can only be effective if it is haphazard, in which case there is no point to it.[24]

What do the macroeconomic models teach us about the formulation of labour market models? The following points seem to emerge as of central importance:

(i) It is vital to distinguish between rationed and unrationed agents. The latter respond only to prices and to changes in prices, the former are moved also by alterations in the constraints to which they are subject. For unrationed

23 See Sargent and Wallace (1976) and Lucas (1981).

24 But there is also no need for it, as the economy is assumed to be always on its long-run supply curve of output.

agents only prices[25] need be included in demand and supply functions, but the same is not of course true of rationed agents.

(ii) Expectations play an essential role in the demand and supply for labour. This is true of the price expectations which have received most recent attention but is equally true of future quantity constraints.

1.5 THE PHILLIPS CURVE AND THE NAIRU

The original Keynesian model took the money wage level as given and determined output and employment conditional on the money wage. Keynes justified this treatment by arguing that money wage changes were largely reflected in changes in the money price level, and for this reason their consequences were relatively unimportant, equivalent to changes in the money supply. Despite this argument, the lack of a theory of money wage determination was felt to be a lacuna in the Keynesian system. Also, as attention turned to the application of Keynesian theory to the formulation of economic policy, many economists worried that the permanent full employment, which was thought to be possible with Keynesian policies, would lead to a serious problem of wage and price inflation.

In this connection the work of Phillips seemed to elegantly fill the gap in the Keynesian model. Phillips investigated the relationship between unemployment and inflation in the United Kingdom and found that inflation responded positively to a decrease in the rate of unemployment, and that the relationship was non-linear, with equal falls in unemployment implying progressively larger increases in inflation. Closer investigation revealed that the relationship reflected to a considerable extent the regular passage of the economy through the trade cycle, with the consequence that deviations from the fitted curve were far from being random, but were highly serially correlated. Right from the beginning, therefore, there was evidence that the Phillips relationship was real but that it might not prove to be robust if the behaviour of the economy altered, due to changes in policy, or for some other reason.

Such suspicions were confirmed in the late 1960s and the 1970s, when Phillips curves that had previously fitted well broke down in Britain, as they did in other countries. The reason was clear. There had been a rapid acceleration in inflation, not due to normal cyclical fluctuations, and the consequence was an upward shift in inflation expectations. It was soon recognized that to take this effect into account the Phillips curve needed to be inflation-adjusted. What was meant was that the relevant index of inflation was a departure from the previously expected rate of inflation. This was the basis of the Friedman–Phelps inflation-adjusted Phillips curve.

Under adaptive expectations the expected rate of inflation would be a weighted sum of previous rates of inflation, with the heavier weights attaching to more

[25] Allowing that a replacement ratio, for example, is a kind of price.

recent observations. In the simplest case, all the weight would attach to the latest observation, and expected inflation would be last period's inflation rate. In that case the deviation of actual inflation from expected inflation would be simply the change in the rate of inflation, and the Phillips relationship taking into account expectations would relate unemployment to changes in the rate of inflation.

It is logical now to look for a rate of unemployment at which the rate of inflation would not change, and the existence of such a rate has been one of the most influential hypotheses of contemporary macroeconomics. The unemployment rate at which the rate of inflation is constant over time is called the non-accelerating inflation rate of unemployment, always abbreviated to the NAIRU.[26]

What determines the NAIRU? That is clearly a very important question for policy, certainly, and even for the most basic understanding of how the economy operates. Clearly if the economy can spend an indefinite time at the NAIRU, and all important variables, including the rate of inflation would be constant, it corresponds to a some kind of equilibrium, but to what kind of equilibrium? One answer to this question is encapsulated in a piece of terminology – the NAIRU may be identified with the natural rate of unemployment.

Friedman (1968) in an often quoted passage explained the natural rate of employment as:

> "*... the level that would be ground out by the Walrasian system of general equilibrium equations provided there is embedded in them the actual structural characteristics of the labor and commodity markets...*"

It is not obvious what this definition adds to the simple statement that the rate of inflation is constant at the natural rate of unemployment. Presumably the hypothetical Walrasian economy has no government and perhaps a constant money supply (if it has money at all). However, it is allowed frictional unemployment, for example, hence the reference to actual structural characteristics of markets. Clearly what the definition reflects is Friedman's view that, were it not for monetary shocks and government interference, the economy would settle down to a constant rate of inflation and a constant level of unemployment. That may well be the case, but it cannot easily function as a definition.

For Friedman the natural rate retains its importance even if the government does interfere and monetary policy is not optimal. This is so because macroeconomic policy can steer the economy away from the natural rate only at the cost of indefinitely accelerating inflation, or decelerating inflation if the economy is above the natural rate of unemployment. Hence in the long run there is no trade-off between unemployment and the inflation rate. As the standard terminology puts it, the long-run Phillips curve is vertical.

[26] The terminology is not quite logical, but it is too late to change it. As inflation is the rate of change of prices, a change in the rate of inflation is an acceleration of prices, but not necessarily an acceleration of inflation. The NAPRU (the non-accelerating prices rate of unemployment) is what is intended. We shall follow the established usage to avoid confusion.

We do best to set aside the issue of defining and measuring the natural rate of unemployment and ask whether a NAIRU can be estimated, and whether the rate so estimated is stable over time. Then it is up to the individual whether he or she chooses to identify the employment level which corresponds to the NAIRU with the natural rate of employment.[27] Fortunately all estimates of the NAIRU and the natural rate for Britain agree on one point, that the NAIRU has risen considerably since the late 1960s.[28] This is obvious when one considers that unemployment has risen continuously during the period while the inflation rate has risen, then fallen, and eventually settled down at a low level.

Blanchard and Summers (1986) provide an exposition of a version of the hysteresis approach. Under hysteresis an increase in unemployment itself affects the NAIRU. The reason in this case is that wage bargaining takes place between employers and employed workers, the 'insiders'. Employed workers, it is argued, have a motive to set the wage level so that they will not become unemployed, but not particularly to lower it so that the unemployed are pulled into employment. This can result in shocks having a permanent effect, as workers randomly placed in unemployment become 'outsiders'. The conclusion is not very different if, for whatever reason, the long-term unemployed become 'unemployable'.[29]

The hysteresis model is attractive in that it can generate the kind of random walk outcome that has sometimes seemed to characterize unemployment. However, one problem with it as expounded by Blanchard and Summers is that the case is not in effect very different from an already familiar argument according to which European unemployment is due to real wage rigidity (see Bruno and Sachs (1985)). The insiders concept simply provides an explanation of the rigidity of wages.

1.6 SOME FEATURES OF THE DATA

In the present section we review a few prominent features of the data on employment, unemployment, wage rates, etc., for the British economy from 1950 to 1985.[30]

First it is pleasant to be able to report that the labour market reflected the growing prosperity of the economy during this period, with the real wage rising almost every year. The real benefit level paid to a reference unemployed man also rose regularly, and the consequence for the retention ratio was that it showed a slight downward trend (Fig. 1.1). Employment rose in line with population but

[27] This identification is not inescapable. For example, the natural rate of employment could be identified with the level of employment at which unemployment and vacancies are equal. See Knight (1987, pp. 99–101).

[28] See in particular Layard *et al.* (1984).

[29] There is evidence that the probability of re-employment declines with the duration of unemployment.

[30] I am grateful to Peter Warburton for providing me with the data which he and Michael Beenstock assembled for their research.

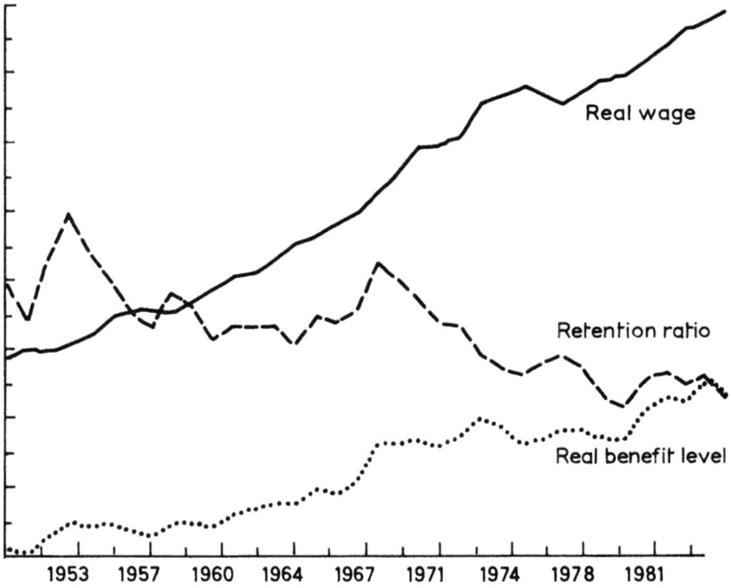
Real wage
Retention ratio
Real benefit level
1953
1957
1960
1964
1967
1971
1974
1978
1981

Fig. 1.1

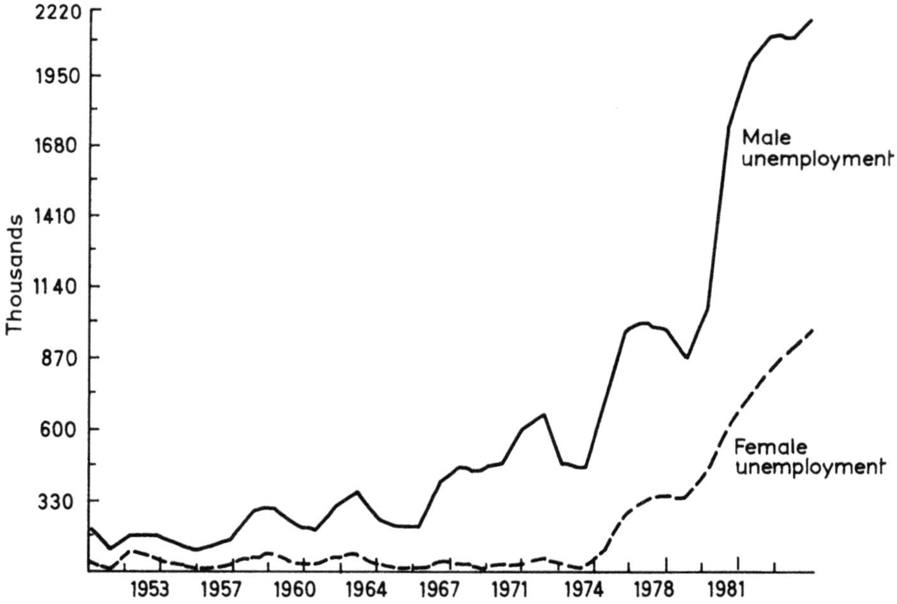
2220
1950
1680
1410
1140
870
600
330
Thousands
Male unemployment
Female unemployment
1953
1957
1960
1964
1967
1971
1974
1978
1981

Fig. 1.2

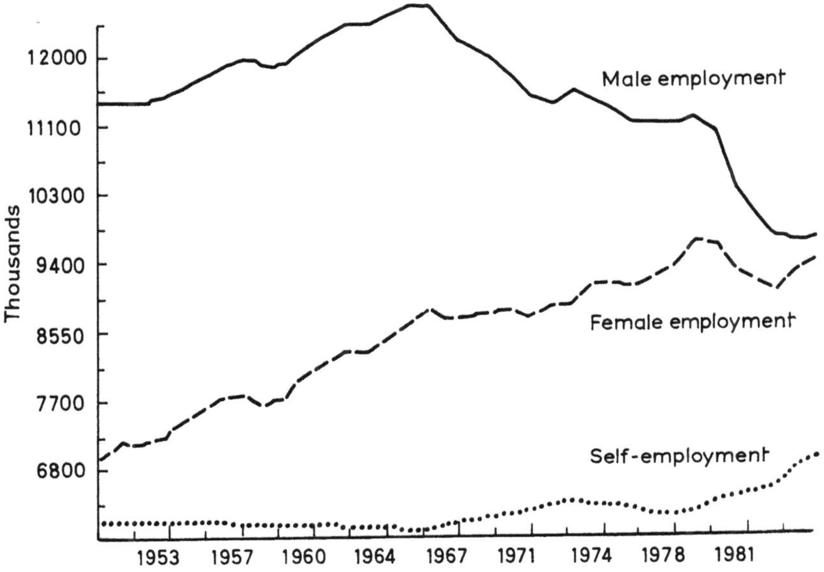
12000
11100
10300
9400
8550
7700
6800
Thousands
Male employment
Female employment
Self-employment
1953 1957 1960 1964 1967 1971 1974 1978 1981

Fig. 1.3

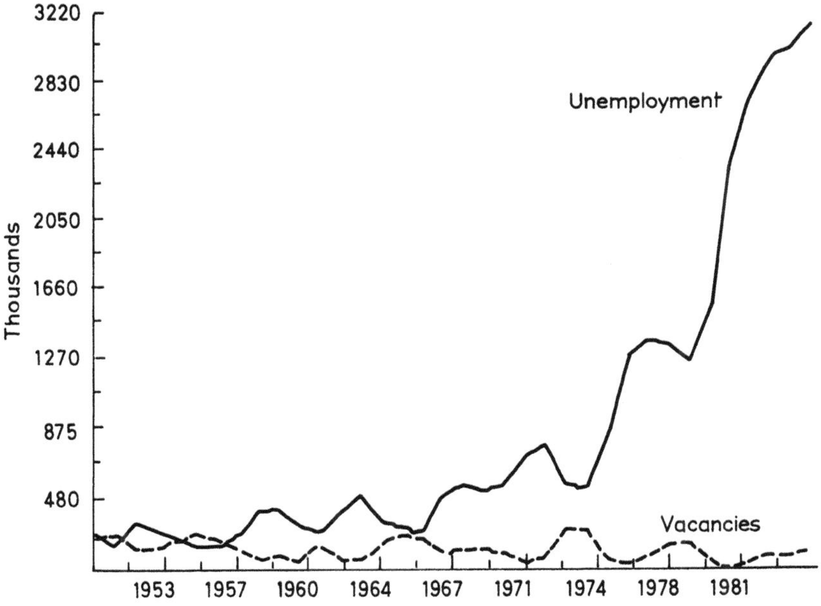
3220
2830
2440
2050
1660
1270
875
480
Thousands
Unemployment
Vacancies
1953 1957 1960 1964 1967 1971 1974 1978 1981

Fig. 1.4

was not closely tied to the population of working age for two reasons. First, unemployment showed an upward trend noticeable throughout the period but becoming very marked towards the end (Fig. 1.2). Secondly, female participation increased throughout the period. In fact male employment was lower in 1985 than it had been in 1950 and the increase in employment over the whole period was entirely accounted for by female employment and self-employment (Fig. 1.3). Figure 1.2 shows that as they participated more in employment females began to share the chief occupational disease of employment which was unemployment.

Figure 1.4 shows total unemployment together with vacancies. It is well known that vacancies underestimate true openings for employment but in the first half of the period vacancies were closely related to unemployment and the relationship was a negative one – more unemployment meant fewer vacancies. Although vacancies were low in the last years of the period their relationship to unemployment broke down from the late 1960s, after which vacancies were much higher than the level of unemployment would have predicted on the basis of previous experience. This was associated in time with the break down of fitted Phillips curve relationships which began to underpredict the rate of inflation at about the same time.

These are not the only features of the data to which attention could be directed, but they do define some features on which a successful model of the labour market would seek to throw light.

REFERENCES

Azariadis, C. (1981) Implicit contracts and related topics: a survey, in *The Economics of the Labour Market* (eds Z. Hornstein *et al.*), HMSO, London.

Beenstock, M. and Warburton, P. (1982) An aggregative model of the UK labour market, *Oxford Economic Papers*, No. 2, July, 253–75.

Blanchard, O. and Summers, L. W. (1986) Hysteresis and European unemployment, in *NBER Macroeconomics Annual* (ed. S. Fischer), MIT Press.

Bliss, C. and Stern, N. H. (1978) Productivity, wages and nutrition; Part I: the theory; Part II; some observations, *Journal of Development Economics*, 5, 363–98.

Bruno, M. and Sachs, J. (1985) *The Economics of Worldwide Stagflation*, Basil Blackwell, Oxford.

Deaton, D. (1982) Employers' demand for labour, in *The Economics of Labour* (eds J. Creedy and R. B. Thomas), Butterworth, London.

Fisher, S. (1977) Long-term contracts, rational expectations and the optimal money supply rule, *Journal of Political Economy*, 85, 191–205.

Friedman, M. (1968) The role of monetary policy, *American Economic Review*, **LVIII**, 1–17.

Friedman, M. (1969) *The Optimum Quantity of Money and Other Essays*, Macmillan, London.

Greenhalgh, C. (1980) Participation and hours of work for married women in Great Britain. *Oxford Economic Papers*, **32**, No. 2, 296–318.

Grossman, S. J. and Hart, O. (1981) Implicit contracts, moral hazard and unemployment, *American Economic Review*, 71(2), 301–7.

Keynes, J. M. (1936) *General Theory of Employment, Interest and Money*, Macmillan, London.
Knight, K. G. (1987) *Unemployment: An Economic Analysis*, Croom Helm, London.
Layard, R. (1986) *How to Beat Unemployment*, Oxford University Press, Oxford.
Layard, P. R. G., Basevi, G., Blanchard, O., Buiter, W. H. and Dornbusch, R. (1984) Europe: the case for unsustainable growth, Centre for European Policy Studies, Paper 8/9.
Lippman, S. and McCall, J. (1976) The economics of job search, *Economic Inequality*, **14**(2), 155–89.
Lucas, R. (1981) *Studies in Business-cycle Theory*, Basil Blackwell, Oxford.
Macdonald, I. M. and Solow, R. M. (1981) Wage bargaining and employment, *American Economic Review*, **71**, 896–908.
Malinvaud, E. (1980) *Theory of Unemployment Reconsidered*, Basil Blackwell, Oxford.
Oswald, A. J. (1982) Wages, trade unions and unemployment. What can simple models tell us? *Oxford Economic Papers*, **34**, 526–45.
Phelps, E. S. (1970) Money wage dynamics and labor market equilibrium, in *Microeconomic foundations of employment and inflation theory* (ed. E. Phelps), Macmillan, London.
Sargent, T. J. and Wallace, N. (1976) Rational expectations and the theory of economic policy, *Journal of Monetary Economics*, **2**(2), 169–83.
Shapiro, C. and Stiglitz, J. E. (1984) Equilibrium unemployment as a worker discipline device, *American Economic Review*, **74**(3), 433–44.
Solow, R. M. (1980) On theories of unemployment, *American Economic Review*, **70**(1), 1–11.
Stiglitz, J. E. (1984) Theories of wage rigidity, paper presented at a conference on Keynes' economic legacy (unpublished).
Varian, H. R. (1978) *Microeconomic Analysis* 2nd edn, W. W. Norton and Co. Inc., New York.
Weiss, A. (1980) Job queues and layoffs in labor markets with flexible wages, *Journal of Political Economy*, **88**(3), 526–38.

2

Some formal models of the aggregate labour market

MARTYN ANDREWS

2.1 INTRODUCTION

The remainder of this book is concerned with the different approaches found in formal, econometric based models of the aggregate labour market. In this chapter we set down a simple taxonomy of models typically found in empirical work, which provides a basic reference for the remaining chapters. Chapters 3 to 6 represent applications for the UK; and finally Chapter 7 presents a selective non-UK view, primarily to provide a contrast with the main substance of this book.

The UK applications described later constitute a representative sample of the recent proliferation in the modelling of labour market aggregates using time-series data. One characteristic of this research activity has been the emphasis on the *joint* determination of these variables, which for the moment we loosely refer to as employment, unemployment, and wages. Thus the models themselves form simultaneous equations systems, which are typically linear and dynamic.[1] Such models are referred to as 'structural', consisting of two or three equations, with each equation corresponding to the actions of a group of similar agents. Thus in the context of the aggregate labour market, the first equation, representing the behaviour of firms, is the demand for labour schedule, although it is often referred to as an employment equation. The second usually represents the behaviour of suppliers of labour, or unions, and is referred to as the supply/wage equation, the latter label occurring because the equation is often renormalized on wages. Sometimes it is estimated as a reduced-form equation for wages. Finally, if the model is to determine unemployment as well as wages and employment, a third structural equation or identity will be present.

The differences between the models that we shall consider are most easily characterized by the assumptions made concerning market structure, namely that

[1] The appropriate acronyms are LDSEM for the Linear Dynamic Simultaneous Equations Model, and DSEM for the more general case.

the markets of interest may be competitive, imperfectly competitive, monopolistic, and so forth. This then bears upon the assumed behaviour for both wages and prices, which are determined, loosely speaking, in the markets for labour and output. In what follows we find it useful to distinguish between the different labour market paradigms by the characterization of both wages and prices as 'fix' or 'flex'. In general, for different paradigms different outcomes for wages and quantity of labour traded occur, as well as there being different policy implications, particularly relating to how any excess supply of labour might be dealt with. Both of these features are reflected in the specifications for the so-called wage/supply and employment/demand equations that are estimated. This is particularly true of the former. However, not all differences in specification arise from these considerations. Thus we start with a discussion of labour demand specifications, which in some ways are more robust to the nature of the model of which it is a part, than is the wage/supply equation.

2.2 LABOUR DEMAND

In what follows we set down in schematic form the kind of labour demand equations typically estimated, first leaving aside the dynamics that arise from considerations such as capital markets, expectations, and adjustment costs. The most general description of the firm's employment decision is written:

$$n^{\mathrm{d}} = n^{\mathrm{d}}(w, Z^{\mathrm{d}}). \tag{2.1}$$

n is the logarithm of the labour input, measured either in heads or man-hours, and Z^{d} represents all variables, endogenous or exogenous, that influence the firm's decision apart from w, the logarithm of the nominal wage. (The superscript 'd' on the dependent variable reflects the possibility that the firm's plans may be frustrated.) More specific forms are generated by considering the market organization of the labour and product markets the firm is buying and selling in, its objectives, the form of its technology, and so forth. Some of the more common specifications are now considered.

We first assume that the firm's output (y) is produced by capital ($\bar{k}$) and labour (n) only, where the former is assumed predetermined (indicated by the overbar) over the run in question: $y = f(n, \bar{k})$. (y and $\bar{k}$ are also logarithmic variables.) The first-order marginal productivity condition for profit maximization, $f_{\mathrm{n}}(n, \bar{k}) = w - p$, describes a negative relationship between the labour input and the logarithm of the real wage, $w - p$, where p is the logarithm of the firm's output price. Rearranged, this gives the following labour demand schedule

$$n^{\mathrm{d}} = n^{\mathrm{d}}(w - p, \bar{k}) \tag{2.2}$$

where $\partial n^{\mathrm{d}}/\partial(w - p) < 0$, and $\partial n^{\mathrm{d}}/\partial k > 0$, the latter being unity under constant returns to scale (CRS). This log-linear formulation can be derived from either a Cobb–Douglas or a constant elasticity of substitution (CES) production function (see Appendix A), the former leading to:

$$n^{\mathrm{d}} = \text{const.} - (1-\alpha)^{-1}(w-p) + \beta(1-\alpha)^{-1}\bar{k}. \tag{2.3}$$

For many years this equation was considered unsuitable for estimation because the capital stock data were thought to be unreliable. Instead $\bar{k}$ was substituted out between the two equations, giving

$$n^{\mathrm{d}} = n^{\mathrm{d}}(w-p, y), \tag{2.4}$$

an equation widely estimated until the late 1970s. However, it is not fully appreciated that to label Equation 2.4 a labour demand schedule is somewhat inappropriate, as is the interpretation of $\partial n^{\mathrm{d}}/\partial(w-p)$ as the slope of the labour demand schedule, which is correctly given by $\partial n^{\mathrm{d}}/\partial(w-p)$ in Equation 2.2. For a CES production function Equation 2.4 is written

$$n^{\mathrm{d}} - y = \text{const.} - s(w-p) + [(1-s)(1-v)/v]y \tag{2.5}$$

where s is the elasticity of substitution between capital and labour and v measures returns to scale. For the Cobb–Douglas case, which specializes a CES production function by having $s=1$, the equation simply says that labour's share, $n^{\mathrm{d}} - y + w - p$, is constant. Comparison with Equation 2.3 illustrates that the coefficients on the real wage are not the same, and that a unit coefficient on output in Equation 2.5 does not necessarily imply constant returns to scale, (i.e. when $s=1$, $v \neq 1$). Thus the best way of interpreting Equation 2.4 is as a rearranged marginal productivity for labour condition for a CES production function without CRS. Constant returns may be tested in the usual way providing $s \neq 1$, delivering a coefficient on y of unity. Notice that because s is the elasticity of substitution of labour with an unspecified factor, data for the second factor is not used unless an instrumental variable (IV) estimator is (properly) employed. Most of these points are to be found in Lucas and Rappings' (1969, Section 3) excellent early discussion of Equation (2.4). Also notice that the equation reduces to a technical relationship between employment and output when the real wage is absent, that is when $s=0$, namely $n=(1/v)y$, which of course is consistent with fixed-coefficient technology. It is this (unrealistic) assumption that implicitly underpins the strand of employment functions that specify a simple autoregressive-distributed lag relationship between employment and output, see Hazledine (1983) for a recent survey.

Further models are generated by introducing a second non-predetermined factor of production, that is $y=f(n,m,\bar{k})$, where m is the logarithm of raw materials and fuel. The motivation behind this possibility was that, because of the impact of OPEC1 and OPEC2 on the price of this variable, its omission might lead to misspecified models when using data for the mid-1970s onwards. Now we may consider the firm's problem under an output constraint, which is often justified at the aggregate level in a 'disequilibrium' model such as (E) (Section 2.3.5) below. Cost-minimization leads to a labour demand schedule of the form

$$n^{\mathrm{d}} = n^{\mathrm{d}}(w-q, \bar{y}, \bar{k}), \tag{2.6}$$

where q is the logarithm of the nominal price of raw materials and fuel. A log-linear version, derived from a Cobb–Douglas production function, is given in Appendix A. This equation is known as the *conditional* labour demand schedule because the firm's employment decision is based on the *relative* cost of the two variable factors, given a predetermined level of output $\bar{y}$. Because the firm's problem is simply to find the optimal position on a given isoquant, it follows that the effect of this variable is unambiguously negative, and so $\partial n^{d}/\partial(w-q)<0$. At this juncture the *real* wage has no bearing on the firm's employment decision. By contrast the *unconditional* labour demand schedule arises if the firm is assumed to maximize profits, leading to:

$$n^{d}=n^{d}(w-p,q-p,\bar{k}). \tag{2.7}$$

The partial derivative $\partial n^{d}/\partial(w-p)$ is unambiguously negative. However, the sign of $\partial n^{d}/\partial(q-p)$ depends on whether the output effect dominates the substitution effect. As noted above, an increase in $q-p$ leads to an increase in labour demand through a substitution effect around a *given* isoquant. However, increased costs at the margin lower both output and employment. Also CRS implies that $\partial n^{d}/\partial\bar{k}=1$. This fourth model, and the less general Equation 2.2, have enjoyed considerable popularity in recent years, particularly in the UK, now that capital stock data are after all used in empirical work, with more and more models being estimated with annual data.

In comparing Equations 2.2 and 2.7 it is clear that the exclusion of raw materials and fuel from the production function means that its price (in real terms) does not appear in the labour demand equation. However, Equation 2.2 is consistent with the more general production function under certain conditions. Firstly output should be measured in value-added, rather than gross, terms (i.e. GDP for the whole economy); and secondly the production function should be separable in raw materials and fuel. (See Appendix A, or for a fuller treatment Bruno and Sachs (1985).)

All the models considered so far have assumed that wages and prices are given to the firm. We now move away from perfect competition and consider what happens when the firm faces a downward sloping demand for its output schedule, $Y(P)$. The firm's problem, if maximizing profits, is

$$\max_{Y} P(Y)Y-Wf^{-1}(Y),$$

where Y depends on $N,\bar{K},\ldots,$ as usual. (All upper case variables are defined naturally.) The first-order condition is known as the marginal revenue product (MRP) schedule:

$$(1-1/\eta)P=W/f_{N}(N,\bar{K},.).$$

where η is the elasticity of demand for the firm's output, $(\partial Y/\partial P)(P/Y)$. The above condition simply equates marginal revenue with marginal cost, but here marginal revenue contains a component $1-1/\eta$, whose inverse is one plus the mark-up, Φ, of price on marginal cost; that is $\Phi = (1 - 1/\eta)^{-1}$. For the special case of perfect competition we have $1/\eta = 0$, or $\Phi = 1$, and so price equals marginal cost. Clearly the MRP schedule may be rearranged so that the labour input is the dependent variable, which can be written in general logarithmic form as

$$n^{d} = n^{d}[w - p + \varphi(\sigma), k] \quad \varphi' < 0 \tag{2.8}$$

($\varphi = \log \Phi$). This equation forms the basis of the imperfectly competitive story described later in model (F) (Section 2.3.6) in this chapter, and subsequently in Chapter 6. (See also Layard and Nickell, 1986.) In general there is no reason why η, and therefore φ, should be constant, and here it is specified as a negative function of σ, aggregate demand. The equation demonstrates that *ceteris paribus* the real wage elasticity remains unaltered irrespective of whether $\varphi(\sigma)$ is included. All that happens in the special case is that prices are exogenous, and that ordinary least squares (OLS) may be used at the firm level of aggregation. Also the mark-up of price on marginal cost is an important determinant of employment, and if prices do not rise as much as marginal cost in response to an increase in demand, employment will rise as well. It is this channel that distinguishes this model from the competitive special case, and consequently the search for the significant presence for aggregate demand variables is an important aspect of their empirical implementation.

It should be stressed that Equation 2.8 is *not* a labour demand equation for the same reason that Equation 2.4 is not: they are both rearranged first-order conditions. The labour demand schedule corresponding to the MRP condition above follows from recognizing that $P = P[Y(N, \bar{K}, .)]$, from which employment can be *solved* in terms of exogenous variables,

$$n^{d} = n^{d}[w, \varphi(\sigma), \bar{k}]. \tag{2.8'}$$

Examples of both Equations 2.8 and 2.8′ are given in Appendix A.

The above equations represent static forms of some of the types of employment/labour demand schedules found in empirical work. However, the equations are usually estimated in dynamic form, almost always specifying lagged employment terms, if not also some distributed lag formulation for the right-hand-side variables. A single lagged dependent variable typically represents sluggish, costly adjustment towards the optimal level of labour input. Further lags may be included if the firm employs different types of worker characterised by different costs of adjustment. More important is the distinction between the stock of a factor (men) and its utilization rate (hours), since one distinguishing characteristic is their relative cost of adjustment, suggesting that they should not be assumed perfect substitutes, at least in the short run. As we shall see this is not taken on board at all in the empirical work discussed in this book, although a strand to the empirical literature does exist; see Nadiri and Rosen (1969) for example. Indeed one important explanation of inter-country differences in employment behaviour is the degree to which non-wage labour costs vary, depending on the particular institutional and legislative environment; see Hart (1984).

The role of expectations is now central to the dynamic specification of empirical labour-market models. For labour demand Sargent's (1978) analysis remains a fixed point, where in the spirit of the competitive paradigm with perfect capital markets, the firm's employment decision depends on an infinite number of relative price terms. Typically these are (implicitly) substituted out, leading to some distributed lag formulation, on the basis of some optimal forecasting scheme; alternatively a finite number may be retained as regressor variables, whereupon some explicit 'rational' expectations hypothesis may be invoked. Finally, we conclude this section by emphasizing that the wage variable appropriate for a labour demand equation is the cost to the firm of hiring labour, including any taxes it pays, which henceforth we denote as w^{p}. Also defining its output or wholesale price as p, we now distinguish the logarithm of the so-called real *product* wage $W^{\mathrm{p}} = w^{\mathrm{p}} - p$, from the pre-tax real wage, $w - p^{\mathrm{r}}$, and the so-called real *consumption* wage, $W^{\mathrm{c}} = w^{\mathrm{c}} - p^{\mathrm{r}}$, where p^{r} is the retail price index and w^{c} the worker's wage net of direct tax. It is easily shown that the difference between the logarithms of the real product and consumption wages is the tax-wedge: $T = t_1 + t_2 + t_3 + s'(p^{\mathrm{m}} - p) + \text{constant}$, where t_1 is the tax-rate on firms; t_2 the direct tax-rate; t_3 the indirect tax-rate; p^{m} the logarithm of the price of imports; and s' the share of imports in value added.

2.3 SOME FORMAL MODELS

We turn to a description of the different approaches to modelling the aggregate labour-market. We start with the most well known of these models, which comprises a labour demand/employment equation, a Phillips curve/wage adjustment equation, and a labour supply equation. In the simple taxonomy that we adopt this is referred to as the 'standard model', an example of which is described in Chapter 4.

2.3.1 The standard model (A)

This paradigm until recently enjoyed almost a complete monopoly in labour market modelling. This derives from its generality, which allows for a wide range of flexibility in the precise specification of the three schedules. Moreover only one assumption about market behaviour is usually made, that employment and labour demand coincide. Thus in the traditional approach, we replace the dependent variable of Equation 2.1 by employment, n. (Throughout the remainder of this chapter this assumption is retained, unless stated otherwise.) Of those outlined in the previous section, the most common specification is Equation 2.4. The equation for wages invariably takes the form of a simple expectations-augmented Phillips curve (APC):

$$\dot{w} = f(\dot{p}, U, Z^{\mathrm{w}}) \qquad 1 > f_1 > 0, f_2 < 0. \tag{2.9}$$

Although this is the standard representation, there are many variants. For example, wages need not be written in derivative form; $\dot{p}$ is inflation, but could be an expectation. Moreover, its coefficient might well be unity, which then defines the so-called accelerationist wage equation. U is unemployment, included as a (poor) proxy for labour-market slackness, but could be the unemployment rate (u^r). A comprehensive survey is that of Santomero and Seater (1978). It is generally accepted that the Phillips curve provides an inadequate description of the data, which is often thought to be symptomatic of its *ad hoc* underpinnings. Subsequent formalization and *ex post* rationalization has led to more structured forms of wage equation, examples of which recur throughout this book. Although there are now more satisfactory specifications, often the data is unable to discriminate between competing theories, and so the Phillips curve in its original form remains popular, particularly when investigating issues such as nominal/real wage rigidity, as discussed later in this chapter.

Assuming the price level, or inflation, is determined outside the labour market, the market itself is closed in one of two ways. The first assumes that the labour force (L) is exogenous, in which case the determination of unemployment and employment can be thought of as opposite sides of the same coin: $U \equiv \bar{L} - N$ or $u^r \approx \bar{l} - n$, where $l = \log(L)$. The second adds a labour force or participation equation:

$$l = l(w^c - p^r, \text{POP}, Z^l). \qquad (2.10)$$

Z^l is a vector of predetermined labour supply variables, including non-labour income and wealth. POP, the logarithm of the population of working age, is identified separately because if its coefficient is unity the dependent variable of Equation 2.10 becomes the participation ratio. The participation decision follows from standard labour supply theory (Killingsworth (1983) provides a good survey) and $w^c - p^r$, the worker's real wage rate, is included as it is the slope of the budget constraint. Often $w^c - p^r$ is absent from Equation 2.10, which then lacks an economic interpretation, and so the determinants of the labour force are solely demographic factors, also included in Z^l. When an equation such as Equation 2.10 is estimated, when taken together with Equations 2.1, 2.9 and the standard identity, the model determines employment, unemployment, labour force, and wages, conditional on prices and other endogenous variables specified in Z^d, for example output, Z^w and Z^l. The remaining variables contained in these three vectors are predetermined, and also implicitly contain all the labour-market taxes defined at the end of the previous section, since the wage variables specified in the four equations do not coincide.

Because observed unemployment cannot be interpreted as equilibrium, or said to be at its natural level on the one hand; or cannot be interpreted as involuntary on the other, the standard model is fairly agnostic about any assumptions of market behaviour. In the next few paradigms these concepts are formalized, where also the microfoundations of the models are based on more rigorous maximizing behaviour on the part of agents. As already noted, one distinguishing feature is the respective characterization of wages and prices as 'fix' or 'flex', and we now consider the market-clearing model in which wages and prices are assumed to be fully flexible. The following sub-section can be thought of as a stylized version of Chapter 5.

2.3.2 The market-clearing model (B)

The market-clearing model is closely related to model (A) above, as is illustrated in Phelps' (ed.) famous (1970) volume. In particular the papers of Mortensen (1970) and Phelps (1970) formalize the Phillips curve as a market-clearing concept, but also the volume contains the seminal, and extremely influential, market-clearing model of Lucas and Rapping (1969).

In schematic form a general market-clearing model consists of a labour supply and a labour demand equation. For supply we write

$$n^s = n^s(w^c - p^r, Z^s) \qquad n_1^s \geqslant 0 \tag{2.11}$$

The first argument is the real consumption wage, W^c. Under certain conditions the equation can be rewritten by renormalizing on this term, with unemployment replacing labour supply using the standard identity,

$$W^c = W(U, Z^s) \qquad W_1 \geqslant 0 \tag{2.12}$$

the labour force term being taken up in Z^s, as in Minford (1983) and Chapter 5. When labour supply is measured in heads, theory indicates that $\partial n^s/\partial W$ should be positive and most cross-section evidence suggests it is often close to zero. This corresponds to a high value for $\partial W/\partial U$ in Equation 2.12, but estimates from market-clearing models such as these tend to be small. On the demand side the firm cannot be output constrained, and so Equation 2.1 can be rewritten as

$$n^d = n^d(w^p - p, Z^d) \tag{2.1'}$$

corresponding to Equations 2.2, 2.4 and 2.7 above.

In the Lucas and Rapping model labour supply is measured in man-hours and there is no distinction made between the participation decision and the rate at which hours are supplied per participant; now the supply-wage derivative cannot be signed. Also on the demand side it is assumed that hours and men are perfect substitutes. Because it is assumed that the real wage adjusts instantaneously to clear the market at all points in time, the market-clearing real consumption wage, W^c, is the solution to $n^s(W^c, Z^s) = n^d(W^p, Z^d)$, that is

$$W^c = \Pi(Z^d, Z^s, T) = W^p(Z^d, Z^s, T) - T. \tag{2.13}$$

T is the tax-wedge, which as noted above, is approximately equal to the sum of all the individual labour-market tax rates. Market-clearing man-hours follows from either supply or demand directly. Because market-clearing employment (i.e. heads) is not explicitly determined in this particular set-up a simple expression for (voluntary) unemployment based on comparing the labour force with employment cannot be utilized. Lucas and Rapping assume that unemployment is a function of the gap between the actual real wage and its long-run value, the so-called 'normal real wage', which under adaptive expectations leads to an unemployment equation of the form

$$u^{r} = g(\dot{w} - \dot{p}, \dot{p}) \qquad g_1 < 0, g_2 < 0. \tag{2.14}$$

This, of course, sustains a Phillips curve interpretation, and is closer to such than Equation 2.12 above, where it is the *level* of real wages that is positively related to unemployment. (The derivative form occurs because adaptive expectations are assumed in forecasting the normal real wage; for a rational expectations treatment see Altonji (1982); Alogoskoufis (1987); Andrews (1987b).) This particular model of voluntary unemployment follows from a two-period model of labour supply, where agents perform an intertemporal maximization problem (see Sargent (1979, ch. 16) for a more rigorous derivation). Because agents are unconstrained, *most* unemployment is voluntary. In contrast to Lucas–Rapping, the Minford model measures employment in heads, and so unemployment enters the model via $N^{s} = N \equiv L - U$. One consequence of this is that *all* unemployment is voluntary, since N^{s} coincides with N, the condition that generates Equation 2.12 from Equation 2.11.

As is well known, the characterization of unemployment as 'voluntary' or otherwise has important implications, as in the former case demand management policy is unable to alleviate the 'problem', should unemployment be considered one. In direct contrast is the standard disequilibrium model, where because the real wage may not be fully flexible, supply and demand are not necessarily equal, leading to the concept of involuntary unemployment. In the first instance this is analysed in a single market framework.

2.3.3 Disequilibrium: single market, model (C)

This class of model historically followed on from (B) above since in Rosen and Quandt's seminal (1978) paper the authors took (a simplified) version of Lucas and Rapping's original specification, but relaxed the market-clearing assumption. Thus Equations 2.11 and 2.1′ are part of the structure. Although there is no necessity for n^{s} to be equal to n^{d}, it is assumed that the real wage in the long run adjusts to clear the market, leading to the following real wage adjustment equation

$$\dot{w} - \dot{p} = \gamma^{-1}(n^{d} - n^{s}) \qquad \gamma > 0. \tag{2.15}$$

By restricting γ to be positive the real wage will move towards its market-clearing level, attained when its rate of change is zero. Substitution of Equations 2.11 and 2.1′ into Equation 2.15 gives the dynamic reduced-form for the real wage:

$$\gamma(\dot{w} - \dot{p}) = n^{d}(w - p, Z^{d}) - n^{s}(w - p, Z^{s}). \tag{2.15′}$$

where for convenience the real wage, $w - p$, is measured in pre-tax terms, any taxes being taken up in Z^d and Z^s. It is easily shown that $\partial(\dot{w}-\dot{p})/\partial(w-p) = (n_1^d - n_1^s)/\gamma$ is negative, ensuring dynamic convergence if the slopes of supply and demand take their usual form. When $\gamma = 0$ the model reduces to market-clearing, since equivalently $n^s = n^d$. On the other hand complete non-market-clearing occurs when $1/\gamma = 0$, or equivalently when $\dot{w} - \dot{p} = 0$. In this latter case real wages are fixed, or exogenously determined, by either the authorities or Trades Unions. Thus empirical estimates of γ are central to tests of market-clearing, where it is noted that it can be reparameterized in terms of a partial adjustment parameter which lies in the [0, 1] interval (Bowden, 1978; Quandt, 1982).

Since employment does not necessarily coincide with either n^s or n^d, further assumptions are required for its determination. For simpler models, as in Rosen and Quandt's, it is in fact determined by the minimum of the two, i.e. $n = \min(n^s, n^d)$, so that all possible (n, W) pairs lie on a 'kink'. This well-known condition arises from the principle of no involuntary exchange, which is a form of rationing scheme. Recently more sophisticated models have been examined which smooth this kink, following Hansen's earlier (1970) suggestion. Instead of there being one aggregate labour market, there are a large number of micro-markets (indexed i), some in excess supply and some in excess demand. The *aggregate* condition above is replaced by one for each micro-market. Aggregate employment is determined by integration (adding-up) if a density function is assumed for the distribution of excess supply over micro-markets. Aggregate employment equations of differing non-linear functional form are generated depending on the choice of density. For example a normal density for $(N^s - N^d)_i$ is used by Andrews and Nickell (1986), for the UK; whereas Lambert (1984) assumes that $(N^s/N^d)_i$ is log-normally distributed.

Irrespective of how employment is determined, these models are notoriously difficult to implement empirically. There are a number of reasons. First, at least one of supply or demand is not observed, which means that the classification of the sample into regime is endogenous to the model, and might not be straightforward in the absence of exogenous data that facilitate the choice. Moreover, the models are non-linear, which requires maximum-likelihood estimation, and it is known that the likelihood function sometimes has peculiar properties. However, some easier-to-use two-stage procedures have been proposed. (Quandt (1982) and Maddala (1983) provide good surveys.) Also note that if an excess supply regime is imposed exogenously *throughout* the sample, i.e. $n^d = n$, all of the above problems disappear, taking us back to model (A) again.

Nonetheless the appeal of the disequilibrium model is that it can indicate the degree to which there is *genuine* labour market tightness over the sample period used, which then should be positively correlated with observed unemployment. (Rosen and Quandt's original attempt did not fare too well; others have been more successful.) Notice also that the wage adjustment equation bears a close resemblance to the Phillips curve, now giving it a non-competitive interpretation. This in turn suggests that this type of excess supply measure might improve the empirical performance of the Phillips curve discussed in model (A). However, the wage adjustment equation above is still *ad hoc*. Finally it should be recognized that there is a potential identification problem, since estimation of three schedules is required of the data for a model explaining two endogenous variables, representing the behaviour of two sets of agents. Consequently other more plausible and explicit accounts of non-competitive behaviour, easier to estimate, are often considered, as discussed in model (F) below. None the less a successful application of the disequilibrium approach, for the UK, is set out in Chapter 3.

The next paradigm is a special case of the standard Rosen and Quandt model in so far as it assumes an excess supply regime throughout. Thus the real wage is restricted to being above, or equal to, its market-clearing level. Model (D) is of interest as it attempts to quantify this degree of labour market tightness empirically, where because of the above assumption, the model is straightforward to estimate. Unlike model (C) it is usually presumed that the price level is explicitly fully flexible, so that the goods market always clears.

2.3.4 The wage-gap model (D)

This methodology was developed within the standard framework of model (A) to examine the hypothesis that unemployment is 'caused by' (or more correctly 'associated with') high levels of real wages. As Sachs notes:

> "*the real wage issue seems to invite extreme positions. In some models real wages are of no significance for macroeconomic outcomes* [*i.e. (B) above*], *while in others they are of decisive importance* [*i.e. (C) above*]. *Not surprisingly, a middle position seems to be supported by the data.*"
>
> (Sachs, 1983, p. 256)

This approach, popularized by Bruno and Sachs (see in particular Bruno and Sachs, 1985), attempts to *quantify* where between the two extremes a particular economy lies, by measuring the gap between the observed real wage and its market-clearing or full-employment level. Thus given a time-series of wage-gap estimates, the empirical relationship between unemployment and high real wages may be examined, which is important both in understanding the operation of a particular economy's labour market, and in designing policies to alleviate unemployment. Clearly there is an inter-country dimension to this analysis, which is discussed more fully in Chapter 7. In the remainder of this sub-section, we explain how such measures are constructed.

The difference between the observed real wage and its full-employment level (or synonomously market-clearing or Walrasian levels), W^{f}, is the wage-gap W^{g}, given by[2]

$$W^{g} = \log (W/P)^{f} - \log (W/P). \qquad (2.16)$$

[2] This use of terminology may be confusing, the wage-gap often referring to the union/non-union wage differential in the labour literature.

All the following methodological points relate to how $(W/P)^f$, a latent variable, is calculated. First, note that a positive wage gap implies that employment is demand determined, and so both W/P and $(W/P)^f$ relate to the real product wage. The central assumption is that the real wage equates to the marginal product of labour at full-employment, that is $(W/P)^f = (\partial Y/\partial N)^f$. Of course this itself is not observed, since the data directly measure average, not marginal, productivity, and moreover full-employment is not always observed. The first problem disappears if a Cobb–Douglas technology is assumed, since the marginal and average products are proportional, and so

$$W^g = \log (Y/N)^f - \log (W/P) + \text{constant}.$$

At this point we note that the wage-gap is directly related to the 'cyclically adjusted labour share'. Defining labour's actual share as $S_N = (W/P)/(Y/N)$, when S_N is evaluated at full-employment we have $S_N^f = (W/P)/(Y/N)^f$, and so the wage-gap is also given by

$$W^g = -\log (S_N^f) + \text{constant}.$$

The second problem, that of not being able to observe full-employment, is usually dealt with by assuming that Y/N attains its full-employment level at cyclical peaks. A complete time-series is then generated by interpolation for the intervening years. The resulting measures may then be renormalized to average zero for any period where it is considered that real wages were genuinely at their full-employment level. For example Bruno and Sachs use 1965–9, the most recent period of full-employment *and* low inflation for the countries they consider. One obvious problem with this is that cyclical peaks do not necessarily imply full employment: 1979 is a case in point.

However, the first assumption is the more serious, namely that average and marginal products are proportional, arising from the fact that for the Cobb–Douglas production function the elasticity of substitution between capital and labour, s, is restricted to unity. Adjustments can be made if the more general CES production function is used, where the following formula applies (Bruno, 1986, p. 39):

$$\frac{\partial \log Y}{\partial \log N} \approx \log (Y/N) + \left[\frac{(1-s)/S_N}{s/(1-S_N)}\right] \log (K/Y)$$

When $s = 1$ the formula reduces to the Cobb–Douglas case, whereby an exogenous increase in the capital stock has no effect on either the output/labour ratio, the marginal product of labour, or the real wage, as labour's share is constant. When $s \neq 1$ the relationship between Y/N, Y/K, and labour's marginal product is not so straightforward. For example if $s < 1$, an increase in the capital/output ratio is necessarily associated with an increase in the labour/output ratio and therefore labour's share, other things being equal. If estimates of s are available, the difference between $\log (Y/N)$ and $\partial \log Y/\partial \log N$ – the latter is related to the wage gap directly – may be calculated provided capital stock data are also available.

Given wage gap estimates, standard regression analysis may be used in establishing the presumed positive relationship between real wages and unemployment. Given that any general labour demand function can be evaluated at the full-employment real wage, then corresponding to Equation 2.1′ we have:

$$n^{\mathrm{f}} = n^{\mathrm{d}}[(w^{\mathrm{p}} - p)^{\mathrm{f}}, Z^{\mathrm{df}}].$$

If $n^{\mathrm{d}}(\cdot)$ is linear, then by subtracting Equation 2.1′ $n^{\mathrm{f}} - n$ is positively related to the wage-gap:

$$n^{\mathrm{f}} - n = n^{\mathrm{d}}[W^{\mathrm{g}}, Z^{\mathrm{df}} - Z^{\mathrm{d}}]. \tag{2.17}$$

Using the unemployment rate as the dependent variable Equation 2.17 is easily estimated, although the usual problem of the data being a poor proxy for excess supply of labour recurs. It should also be stressed that any relationship found between labour-market tightness and the wage-gap does not imply any particular causality between the two endogenous variables, and I.V. therefore is the appropriate estimator.

The two non-market-clearing models we have considered thus far are characterized by real wage sluggishness, although the actual cause may be nominal wage sluggishness, depending on what is assumed about price flexibility. Now we consider a class of model which relaxes the assumption that the market for goods always clears. This is characterized as the fix-price/fix-wage model, which is of particular interest in that two different unemployment concepts may be identified, known as 'Keynesian' and 'classical'.

2.3.5 Disequilibrium: multi-market (model E)

There is an important and influential strand in the macroeconomics literature, following Barro and Grossman (1971), which analyses disequilibrium in both the labour and goods markets. This analysis was partly a response to the view that Keynes' contribution to macroeconomics was *not* simply the analysis of some special cases of the classical paradigm. In what follows we select out of this literature (which is associated with Malinvaud, Muellbauer and Portes, *inter alia*), one particular stylized version, that of Sneessens (1983), since it underpins empirical applications for certain European countries as discussed in Chapter 7.

The relevance of the second market may be described in general terms as follows:

> *"Let us consider a macroeconomic model describing the goods and labour markets only. Because prices are sticky, demands and supplies do not usually match, and an allocation procedure (the rationing scheme) is needed to clear the markets. This entails two drastic differences with respect to Walrasian models. One relates to the impact of perceived quantity constraints on the behaviour of producers and consumers (the so-called spill-over effect) and leads to the distinction between effective and notional (or Walrasian) trade offers. The other follows from the usual assumption that transacted quantities only are observed on each market; provided that the*

allocation procedure is market-by-market efficient, this leads to the use of the well-known min condition."

(Sneesens, 1983, p. 194)

Clearly the addition of the second market gives rise to the basic idea of a spillover, and the following simple example illustrates how one occurs and its importance. Suppose there is excess supply in the labour market. This generates labour income lower than its Walrasian level because some suppliers are unable to sell their labour. This results in lower consumption demand on the goods market. Because consumption demand now effectively depends on the quantity of labour traded, the so-called 'effective' demand schedule is written $y^{d}(n;.)$, and is distinguished from the 'notional' (or Walrasian) demand that occurs in the absence of spillover effects, which depends on relative prices only. Suppose also there is excess supply in the goods market, so production is lower than its competitive level. This also spills over into the labour market, and effective labour demand, written $n^{d}(y;.)$, is also lower than its notional counterpart. (Note that this type of labour demand schedule arises from cost minimization, not profit maximization, an example of which is Equation 2.6 above.) In this example both markets are determined on the demand side, and so the two effective demand schedules $n=n^{d}(y;.)$ and $y=y^{d}(n;.)$, when taken together, give a solution for employment and output, in terms of the exogenous wages and prices, below their Walrasian levels.

There are four possible disequilibrium configurations, taking pairwise goods and labour with excess supply and excess demand. Thus excess supply of labour, or involuntary unemployment, can occur with either excess supply or excess demand in the goods market, giving rise to the so-called Keynesian and classical regimes noted above.

The following framework provides a basis for empirical applications:

$$\begin{aligned} y &= y^{s} = f(n) \\ n &= \min(n^{d}, n^{s}) \\ n^{d} &= f^{-1}(\hat{y}^{d}) \quad \text{if} \quad \hat{y}^{d} < y^{f} \\ &= f^{-1}(y^{f}) \quad \text{otherwise} \\ n^{s} &= n^{s}(L). \end{aligned} \tag{2.18}$$

This is less general than the theory model discussed above. In particular the short-run production function given in the first equation replaces the standard $y=\min(y^{s}, y^{d})$ condition, since it is assumed that excess demand for goods may be satisfied by increased production or increased imports, or by running down inventories. Labour demand is determined by the expected level of final demand, $\hat{y}^{d}$, provided it is not greater than potential/Walrasian output, y^{f}. Labour supply is Walrasian in the sense that workers are not constrained on the goods market. The model is now written as $y=f(n)$ plus

$$n = \min[f^{-1}(\hat{y}^{d}), f^{-1}(y^{f}), n^{s}(L)]$$

or

$$n = \min\,[n^{\mathrm{k}}, n^{\mathrm{f}}, n^{\mathrm{s}}]. \tag{2.19}$$

Of the four possible regimes, the three standard ones found in multi-market disequilibrium models correspond to which of the three arguments is binding. In the first two of these, firms are able to hire as much labour as they wish. If then n^{k} is binding ($n = n^{\mathrm{k}}$) there is excess supply in both markets ($n^{\mathrm{d}} < n^{\mathrm{s}}$ and $\hat{y}^{\mathrm{d}} < y^{\mathrm{f}}$), which defines the Keynesian regime; whereas if a capacity constraint is binding ($n^{\mathrm{d}} < n^{\mathrm{s}}$ and $\hat{y}^{\mathrm{d}} > y^{\mathrm{f}}$), we have the classical regime, that is $n = n^{\mathrm{f}}$. The third case, Repressed Inflation, occurs when firms face labour shortages and is the only occasion where employment and labour demand do not coincide, since now $n = n^{\mathrm{s}}$ is the binding constraint.

Estimation of the model is based around Equation 2.19. Like the Rosen and Quandt model, linear equations may be specified for $y^{\mathrm{d}}(\ )$, $n^{\mathrm{d}}(\ )$, $n^{\mathrm{s}}(\ )$, and $f(n)$, unless n^{s}, for example, is considered exogenous. Moreover, it usually is thought inappropriate that wages and prices are exogenous; equations may be estimated which capture their sluggish movement towards Walrasian equilibrium, as in Equation (2.15) above. More recently disequilibrium models have moved away from the idea that it is wages and prices that lead to constraints on agents behaviour, but rather there are technical constraints to (short-run) expansion in output and employment. In this case the production function $f(n)$ should exhibit some degree of fixity in the short run, as must then the corresponding $n^{\mathrm{d}}(\ )$ schedule. (Recall what happens to Equation 2.5 when $s = 0$.) What *is* certain is that estimation is not straightforward. The problems noted above are compounded when the data have to choose between a larger number of regimes; and also with more equations there are more parameters to estimate simultaneously.

2.3.6 Imperfect competition (F)

Very recently a model based on imperfect competition has been estimated by Layard and Nickell, using UK data, which is discussed in further detail in Chapter 5. In this section we set down in our schematic form the schedules of interest, and note their relationship to models (A) to (E) above.

The motivation for developing a model which is quite different from either the Walrasian or disequilibrium type is that the former often is not taken seriously given the historically high levels of unemployment, all of which cannot be 'voluntary'; and the fix-price fix-wage assumptions required to generate 'involuntary' unemployment appear *ad hoc*, or implausible, for the latter. We let Layard and Nickell take up the issue:

> *"The labour demand function that we use cuts through the fruitless debate now raging (especially in Europe) as to whether current unemployment is 'classical' or 'Keynesian'.... The whole debate is set in framework of perfect competition... [and so] prices are set by impersonal forces, and it is not clear what could possibly stop prices clearing the market. It is much more*

> *reasonable to think of prices being set by imperfectly competitive firms, existing prices being the best they can think of, given the demand they face. In this context, firms' demand for labour will depend on both the real product wage and the level of real aggregate demand.*"
>
> (Layard and Nickell, 1986, p. 121)

Clearly the type of labour demand schedule appropriate for firms selling in imperfectly competitive product markets is as given in Equation 2.11 above. It is worth repeating that under perfect competition, or under imperfect competition with a *constant* elasticity product demand schedule, the term in aggregate demand, σ, disappears. Since the firm sets price as a mark-up on marginal cost, it is clear that when Equation 2.11 is rearranged so that $p-w$ is the dependent variable, it may be described as a price equation. However, this argument, that the price and labour demand schedules are one and the same, is rather too stylized, and does not carry through in more realistic applications where there are a large number of firms, and where a distinction is made between potential aggregate output and actual aggregate demand.

In order to complete the model a real wage equation is required. As the nature of the forces that determine wages are independent of the type product market in which the firm is selling, the story as far as model (F) is concerned finishes here. However, a synopsis of the wage equations already surveyed in this chapter demonstrates the wide range of possibilities. These are the Phillip's curve, Equation 2.9; the rearranged competitive labour supply schedule, Equation 2.12; the reduced-form competitive wage equation, Equation 2.13; and the disequilibrium wage adjustment Equations 2.15 or 2.15′. Another possibility is now considered, based on the following models.

(a) Bargaining models

These may be derived when firms and unions, the latter acting on behalf of suppliers of labour, are assumed to bargain over wages and possibly employment. There are two sub-cases: when the bargain is over wages *and* employment, and when the firm retains the right to manage. The more general of the two is set out in MacDonald and Solow (1981). Consider real-wage and employment space. When the firm's iso-profit contours are set against the union's utility contours, the locus of tangency points represents all possible combinations where neither side can improve his lot except at the expense of other. This schedule lies to the right of the competitive labour demand schedule, intersecting at the competitive outcome. The position of the final equilibrium, determined by the outcome of a so-called Nash bargain, represents the relative 'power' of one side over the other. As yet there are very few empirical applications using aggregate data.

When the bargain is over wages only, the so-called 'right-to-manage' model, the solution is restricted to being on the *competitive* labour demand schedule, and will lie between two extreme positions. One is where firms are all powerful, and are able to pay the minimum wage possible – the competitive wage – for maximum employment (and profit). At the other extreme the union is all powerful and sets the real wage unilaterally. Any position inbetween represents a solution which is determined by factors that influence *both* the firm and the union in the bargain. These are Z^d and (U, Z^s) respectively. Clearly the solution can be written

$$w - p = f(U, Z^s, Z^d) \qquad f_1 > 0 \tag{2.20}$$

The unemployment term is of some interest, and arises because the union cares about outsiders (who might be potential members) and thus reduces its power in the bargain.

(b) Unions set the wage

This is the possibility noted above, where unions are all powerful in the bargain, known as the 'union-monopoly' model (Dunlop, 1944). In this special case there is no bargaining; firms set employment and unions set the wage.

(c) Firms set the wage

These are the so-called efficiency wage models where the firm sets the wage such that marginal benefit equates with marginal cost, but in particular the former includes lower quit rates, more effort, and so forth, reflecting how 'insiders' have some bearing on the wage set. However, the wage set will also reflect 'outside' opportunities such as alternative wages, unemployment and benefits. Thus Equation 2.20 reflects this type of model as well as the explicit bargaining solution suggesting that tests to discriminate between the two are difficult with aggregate data[3].

In concluding this sub-section on imperfect competition, the real wage equation, Equation 2.20, typically estimated encompasses most of the above possibilities, and is therefore more general than those described earlier. One commendable feature is that its dependent variable is the level of real wages, not the rate of change, in contrast to both Equations 2.9 and 2.15. Moreover, when part of a two-equation system such as Equation 2.8 there are three more attractive features. First the model determines the levels of both employment and the real wage. Second, being linear, it is easy to estimate, in contrast to the disequilibrium models. Finally it encompasses some interesting special cases, notably models with competitive labour and competitive product markets. We now turn to our

[3] More rigorous derivations of these wage equations, and indeed some of the models discussed in this chapter, are given in Nickell (1984). Andrews (1987a) is a survey of empirical applications for the UK, including two models not discussed in this book, namely Andrews (1987b) and Nickell and Andrews (1983). The former is a competitive model more in the spirit of the Lucas and Rapping version of the paradigm than the model set out in Chapter 5 later. The latter is based on the bargaining model set out above, consisting of Equations 2.20 and 2.7. To all intents and purposes this model has been superseded by the Layard and Nickell model set out in Chapter 6, i.e. Equations 2.20 and 2.8.

final model, which is distinctly different from *all* the structural models discussed above, simply because there is no discernible labour market structure.

2.3.7 Non-structural models (G)

In each of the six previous models, the aggregate labour market has been represented by one, two, or three 'structural' equations, each roughly corresponding to the market responses of a particular group of agents. Estimation of these equations is straightforward when the models form *linear* dynamic simultaneous equations models (LDSEMs), either as a complete model by full (FIML), or information maximum likelihood, (3SLS), 3-stage least squares, or equation by equation using I.V. or 2SLS. Usually ordinary least squares (OLS) is inappropriate because of the presence of right-hand-side endogenous regressors, the exception being the type of reduced-form equation such as Equation 2.13. As noted it is disequilibrium models that are intrinsically non-linear in nature, estimation requiring some numerical maximum likelihood optimization procedure, although two-stage methods are available.

We now consider models that are not made up of structural equations, such as those described as labour demand or labour supply. Instead time-series representations of the processes that generate the endogenous variables of interest – employment, unemployment, and wages – are estimated (Sims, 1980). In the vector autoregressive model, VAR, a vector of endogenous variables, which for expositional convenience we take as n and $w-p$, is determined by its own lagged values; and also allows for a autoregressive errors. This we can write as

$$\begin{bmatrix} n_t \\ (w-p)_t \end{bmatrix} = \begin{bmatrix} a_{11}(L) & a_{12}(L) \\ a_{21}(L) & a_{22}(L) \end{bmatrix} \begin{bmatrix} n_{t-1} \\ (w-p)_{t-1} \end{bmatrix} + \begin{bmatrix} u_{1t} \\ u_{2t} \end{bmatrix},$$

and

$$\begin{bmatrix} u_{1t} \\ u_{2t} \end{bmatrix} = \begin{bmatrix} \gamma_{11}(L) & \gamma_{12}(L) \\ \gamma_{21}(L) & \gamma_{22}(L) \end{bmatrix} \begin{bmatrix} u_{1,t-1} \\ u_{2,t-1} \end{bmatrix} + \begin{bmatrix} e_{1t} \\ e_{2t} \end{bmatrix},$$

or in matrix notation as

$$\boldsymbol{y}_t = \boldsymbol{A}(L)\boldsymbol{y}_{t-1} + \boldsymbol{u}_t, \quad \boldsymbol{u}_t = \boldsymbol{\Gamma}(L)\boldsymbol{u}_{t-1} + \boldsymbol{e}_t \tag{2.21}$$

e_{it} is a zero mean, constant variance, stochastic process; L is the lag operator, although 'leads' (L^{-1}) are not ruled out in the $\boldsymbol{A}$ (L) matrix. (The *vector* $\boldsymbol{y}_t$ is not to be confused with the scalar y_t above, real output.)

Within this framework certain hypotheses about the behaviour of employment and/or the real wage may be examined and tested. A flavour of the possibilities is as follows:

(i) Whether the real wage and employment are positively or negatively related, or independent. The importance of ascertaining which of these possibilities is supported by data is of central importance in macroeconomics. The introduction to Geary and Kennan (1982) illustrates (paraphrased for brevity) the degree to which the matter is unresolved:

"The neoclassical and Keynesian theories of employment both predict an inverse relationship between employment and real wages in the short run. The observed correlation ... has usually failed to conform to this prediction. Dunlop (1938) and Tarshis (1939) reported... [a result]... was interpreted as evidence against the theoretical prediction; the statistical significance of these results was, however, challenged by Ruggles (1940) and Tobin (1948). More recent studies by Kuh (1966) and Bodkin (1969) failed to detect a significant relationship between real wages and employment." ... [They conclude] "Neftci (1978) and Sargent (1978), however cast doubt on the empirical conventional wisdom."

(Geary and Kennan, 1982, pp. 854–5)

The stylized fact that the relationship is not always negative motivated the development of the fix-price/fix-wage model of Barro and Grossman (1971);

(ii) Whether employment fluctuates more than the real wage. This stylized fact underpins the development of the MacDonald and Solow (1981) union bargaining model;

(iii) Whether either the nominal or real wage exhibits rigidity. The first possibility is a centrepiece of Keynesian economics, from which the so-called implicit-contract theories have attempted to explain both rigid wages and layoff unemployment as the rational outcome of *explicit* maximizing behaviour (Baily, 1974; Azariadis, 1975). Another explanation is that wage contracts are written in nominal terms, for long periods, 'staggered' from one group of workers to the next. The second possibility, that the real wage is rigid, or follows a random walk, is, of course, the extreme disequilibrium case (Taylor, 1979).

The third hypothesis can be examined by considering the bottom equation, appropriately specified in real or nominal terms, by setting $a_{21}(L) = 0$, and testing whether the polynomial $1 - a_{22}(L)$ contains a unit root. Similarly examination of the cross-polynomials $a_{21}(L)$ and/or $a_{12}(L)$ is central to tests of the independence hypothesis. Although (iii) is one way by which wage ridigity may be examined, there are two others. The first is simply to calculate directly the sample standard deviations of the (log of) wages (real or nominal) or the sample standard deviation of their rates of change. One obvious problem with calculating for example $\widehat{\text{SD}}(\Delta w_t)$, a small value corresponding to a low degree of wage responsiveness, is that the size of the statistic cannot be easily judged as large or small, although comparisons with the same statistic for employment (see (ii) above), or across countries, can be made. The second focuses on the estimation of the standard Phillips curve or wage equation, as given by Equation 2.9 above. There is some confusion as to which parameters actually measure rigidity, although it appears that the lagged wage terms and the coefficient on unemployment are important in this regard.

As an econometric methodology the vector-autoregressive approach is more popular in North America than it is in Europe, where estimation of structural equations appears more prevalent. As one is usually not a special case of the other, the debate of which is the 'better' is still unresolved, and is particularly keen when 'vector autoregressive' and 'structural' large-scale macro-models are compared, their forecasting performance being a particularly important discriminating criterion. For example, moving away from the aggregate labour market for the moment, Wallis (ed.) *et al.* (1986) conclude that, in comparing different types of *ex post* forecast generated by the UK macroeconometric models, the following ranking occurs. The best forecasts are those that include all the 'adjustments' the model proprietors wish to make. Next come those generated by the 'hands-off' approach; which in turn are better than those generated by simple vector-autoregressions for the endogenous variables of interest. Irrespective of this kind of evidence, one point should be emphasized, and is of general applicability in comparing any of the models discussed in this book. When one model is a special case of another, standard tests can be brought to bear; when they are not nested, it is often the case that both contain useful information, and one should not be jettisoned in favour of another. It is hoped that the framework set out in this chapter goes some way to clarifying the relationship between all the empirical models that are described in the rest of this book.

APPENDIX A SOME BACKGROUND NOTES ON LABOUR DEMAND EQUATIONS

For the Cobb–Douglas production function $Y = AN^{\alpha}\bar{K}^{\beta}$, the first-order marginal productivity condition is

$$\partial Y/\partial N = \alpha A N^{\alpha-1}\bar{K}^{\beta} = W/P. \tag{A1}$$

Taking logs and solving for $\log N$ leads to Equation 2.3, corresponding to the general equation (Equation 2.2). For the CES production function

$Y = A[(1-\delta)N^{-\rho} + \delta\bar{K}^{-\rho}]^{-\nu/\rho}$, the first-order marginal productivity condition is

$$\partial Y/\partial N = \nu(1-\delta)\frac{Y^{1+\rho/\nu}}{N^{1+\rho}} = W/P. \tag{A2}$$

where $\bar{K}$ has been substituted out, using Y. By taking logs of (A2) we obtain the rearranged marginal productivity condition given by Equation 2.5 of the main text, specializing the generic form (Equation 2.4). The equivalent Cobb–Douglas case follows by substituting $s^{*}= 1$ into Equation 2.5 of the main text. If, however, we substitute $s = 0$, that is fixed coefficient technology, Equation 2.5 reduces to $n^{d} = \text{const.} + \nu^{-1}y$. Obtaining an equation corresponding to Equation 2.2 based on the CES production function is not easy, although Bruno (1986) presents the following approximation:

$$\log (N^{d}/K) = -(s/S_{k}) \log (W/P) \tag{A3}$$

where S_{k} is capital's share.

Extending the model to two short-run factors, for cost minimization, i.e.

$$\min C = WN + QM - \lambda[f(N, M, \bar{K}) - \bar{Y}]$$

there are three first-order conditions

$$W = \lambda \partial f / \partial N,$$
$$Q = \lambda \partial f / \partial M,$$
$$\bar{Y} = f(N, M, \bar{K}).$$

These solve for N, (M and λ) as given in Equation 2.6 of the main text. For a Cobb–Douglas under CRS, that is $Y = N^{\alpha} M^{\beta} \bar{K}^{1-\alpha-\beta}$, the above three equations specialize to

$$W = \lambda\alpha N^{\alpha-1} M^{\beta} \bar{K}^{1-\alpha-\beta}$$
$$Q = \lambda\beta N^{\alpha} M^{\beta-1} \bar{K}^{1-\alpha-\beta}$$
$$\bar{Y} = N^{\alpha} M^{\beta} \bar{K}^{1-\alpha-\beta}$$

The solution for $\log N$ is

$$\log N = \text{const.} - \frac{\beta}{\alpha+\beta} \log(W/Q) + \frac{\log \bar{Y}}{\alpha+\beta} - \frac{1-\alpha-\beta}{\alpha+\beta} \log \bar{K}.$$

For profit-maximization, $\max \Pi = P \cdot f(N, M, \bar{K}) - WN - QM$, the first-order conditions, written in general and for a Cobb–Douglas, are given by

$$\partial f / \partial N = W/P = \alpha A N^{\alpha-1} M^{\beta} \bar{K}^{1-\alpha-\beta}$$
$$\partial f / \partial M = Q/P = \beta A N^{\alpha} M^{\beta-1} \bar{K}^{1-\alpha-\beta}.$$

Solving for $\log N$ in general gives Equation 2.7 of the main text; for the Cobb–Douglas:

$$\log(N/K) = -\frac{1-\beta}{1-\alpha-\beta} \log(W/P) - \frac{\beta}{1-\alpha-\beta} \log(Q/P) + \text{const.}$$

Notice that the coefficient on $\log(Q/P)$ may be split into the substitution effect $\beta/(\alpha+\beta)$ and the output or scale effect, $-\beta/(1-\alpha-\beta)/(\alpha+\beta)$. Clearly the former dominates, since the total effect $-\beta/(1-\alpha-\beta)$ is negative.

Turning now to imperfect competition, suppose value-added output is produced under constant returns,

$$Y/\bar{K} = (N/\bar{K})^{\alpha}.$$

and the firm faces a demand for its output schedule

$$P = Y^{-1/\eta}.$$

Equating marginal revenue with marginal cost can be written in one of two ways:

$$(1 - 1/\eta) Y^{-1/\eta} = \alpha^{-1} W (Y/\bar{K})^{1/\alpha - 1} \tag{A4}$$

or

$$(1 - 1/\eta) P = \alpha^{-1} W (N/\bar{K})^{1-\alpha} \tag{A5}$$

From (A4) a *solution* for Y, and therefore N, can be obtained in terms of W, $\bar{K}$, η and α, which is the firm's labour *demand* schedule, corresponding to Equation 2.8′ of the main text:

$$\log(N/\bar{K}) = \text{const.} + [\alpha(1-\eta^{-1})-1]^{-1}(\log W - \eta^{-1}\log\bar{K}).$$

However, using (A5), and writing $-\log(1-1/\eta) = \phi(\sigma)$, we get

$$\log(N/\bar{K}) = \text{const.} - (1-\alpha)^{-1}[\log(W/P) + \phi(\sigma)]$$

corresponding to Equation 2.8 of the main text.

Value added

If raw materials and fuel, M, are separable from the other factors of production, then the profit maximization first-order conditions change. Writing the production function as $Y = f(N,\bar{K}) + \lambda(M)$, the firm's problem is to maximize

$$(\Pi/P) = f(N,\bar{K}) + \lambda(M) - (W/P)N - (Q/P)M$$

Solving the first-order conditions leads to:

$$N = N^{d}(W/P,\bar{K})$$
$$M = M^{d}(Q/P)$$

In particular it should be noted that labour demand no longer depends on Q/P, as in Equation 2.7 of the main text. An equivalent way of viewing the same problem is to define $Y^{*}(N)$ as real value-added output, $Y^{*} \equiv Y - (Q/P)M$. These points are easily illustrated using the following production function

$$Y = N^{\alpha}\bar{K}^{\beta} + M^{\gamma},$$

with first-order conditions (A1) above and

$$\partial Y/\partial M = \gamma M^{\gamma-1} = Q/P.$$

REFERENCES

Alogoskoufis, G. S. (1987) Aggregate employment and intertemporal substitution in the UK, *Economic Journal*, 97, 403–15.

Altonji, J. G. (1982) The intertemporal substitution model of labour market fluctuations: an empirical analysis, *Review of Economic Studies*, **49** (supplement), 783–824.

Andrews, M. J. (1987a) Empirical models of the UK aggregate labour market: a survey, *Economic Perspectives*, **4**, 175–223.

Andrews, M. J. (1987b) The aggregate labour market – an empirical investigation into market clearing for the UK, *Economic Journal*, **97**, 157–76.

Andrews, M. J. and Nickell, S. J. (1986) A disaggregated disequilibrium model of the labour market, *Oxford Economic Papers*, **38**, 386–402.

Azariadis, C. (1975) Implicit contracts and underemployment equilibria, *Journal of Political Economy*, **83**, 1183–1202.

Baily, M. N. (1974) Wages and employment under uncertain demand, *Review of Economic Studies*, **41**, 37–50.
Barro, R. J. and Grossman, H. (1971) A general disequilibrium model of income and employment, *American Economic Review*, **61**, 82–93.
Bowden, R. J. (1978) *The Econometrics of Disequilibrium*, North Holland, Amsterdam.
Bruno, M. (1986) Aggregate supply and demand factors in OECD unemployment: an update, *Economica* (supplement), 35–52.
Bruno, M. and Sachs, J. D. (1985) *Economics of Worldwide Stagflation*, Blackwell, Oxford.
Dunlop, J. T. (1938) The movement of real and money wage rates, *Economic Journal*, **48**, 413–34.
Dunlop, J. T. (1944) *Wage Determination Under Trade Unions*, Macmillan, New York.
Geary, P. T. and Kennan, J. (1982) The employment-real wage relationship: an international study, *Journal of Political Economy*, **90**, 854–71.
Hansen, B. (1970) Excess demand, unemployment, vacancies and wages, *Quarterly Journal of Economics*, **84**, 1–23.
Hart, R. A. (1984) *The Economics of Non-wage Labour Costs*, George Allen and Unwin, London.
Killingsworth, M. R. (1983) *Labour Supply*, Cambridge University Press, Cambridge.
Kuh, e. (1966) Unemployment, production functions and effective demand, *Journal of Political Economy*, **74**, 238–349.
Lambert, J. P. (1984) Disequilibrium macro-models based on business survey data: theory and estimation for the Belgian manufacturing sector, Cambridge University Press (forthcoming).
Layard, P. R. G. and Nickell, S. J. (1986) Unemployment in Britain, *Economica*, **53** (supplement), 121–70.
Lucas, R. E. and Rapping, L. A. (1969) Real wages, employment and inflation, *Journal of Political Economy*, **77**, 721–54.
Macdonald, I. M. and Solow, R. M. (1981) Wage bargaining and employment, *American Economic Review*, **71**, 896–908.
Maddala, G. S. (1983) *Limited-dependent and Qualitative Variables in Econometrics*, Cambridge University Press, Cambridge.
Minford, A. P. L. (1983) Labour market equilibrium in an open economy, *Oxford Economic Papers*, **35** (supplement), 207–44.
Mortensen, D. T. (1970) A theory of wage and employment dynamics, in *Micro-economic foundations of employment and inflation theory*, (eds E. S. Phelps *et al.*), Macmillan, London.
Nadiri, M. I. and Rosen, S. (1969) Interrelated factor demand functions, *American Economic Review*, **59**, 457–71.
Nickell, S. J. (1984) The modelling of wages and employment, in *Econometrics and Quantitative Economics*, (eds D. F. Hendry and K. F. Wallis), Basil Blackwell, Oxford.
Nickell, S. J. and Andrews, M. J. (1983) Unions, real wages and employment in Britain 1951–79, *Oxford Economic Papers*, **35** (supplement), 183–220.
Phelps, E. S. (1970) Money wage dynamics and labour market equilibrium, in *Micro-economic foundations of employment and inflation theory*, (eds E. S. Phelps *et al.*), Macmillan, London.
Quandt, R. E. (1982) Econometric disequilibrium models, *Econometric Reviews*, **1**, 1–63.
Rosen, H. S. and Quandt, R. E. (1978) Estimation of a disequilibrium aggregate labour market, *Review of Economics and Statistics*, **60**, 371–79.
Ruggles, R. (1940) The relative movements of real and money wage rates, **55**, 130–49.
Sachs, J. D. (1983) Real wages and unemployment in the OECD countries, *Brookings Papers on Economic Activity* **1**, 255–304.
Santomero, A. M. and Seater, J. J. (1978) The inflation-unemployment trade-off: A critique of the literature *Journal of Economic Literature*, **16**, 499–544.

Sargent, T. J. (1978) Estimation of dynamic labour demand schedules under rational expectations, *Journal of Political Economy*, **86**, 1009–44.
Sargent, T. J. (1979) *Macroeconomic Theory*, Academic Press, New York.
Sims, C. A. (1980) Macroeconomics and reality, *Econometrica*, **48**, 1–48.
Sneessens, H. (1983) A macroeconomic rationing model of the Belgian economy, *European Economic Review*, **20**, 193–215.
Tarshis, L. (1939) Changes in real and money wage rates, *Economic Journal*, **49**, 150–4.
Taylor, J. B. (1979) Staggered wage setting in a macro-model, *American Economic Review*, **69**, 108–13.
Tobin, J. (1948) Money wage rates and employment, in *The New Economics: Keynes' Influence on Theory and Public Policy* (ed. S. E. Harris), Dobson, London.
Wallis, K. F. (ed.) Andrews, M. J., Fisher, P. G., Longbottom, J. A. and Whitley, J. D. (1986) *Models of the UK Economy: A Third Review by the ESRC Macroeconomic Modelling Bureau*, Oxford University Press, Oxford.

3

The disequilibrium approach to modelling the labour market

STEPHEN HALL and BRIAN HENRY

3.1 INTRODUCTION

In this chapter we will explore the application of disequilibrium modelling techniques to the UK labour market. The concept of equilibrium is obviously an important one in economics but it is not entirely unambiguous. Equilibrium is sometimes taken to mean that demand is equal to supply (in all markets if more than one market is being considered); an alternative definition is that the economic system is 'at rest' and so there are no forces tending to bring about change. These two definitions are not identical; we can for example consider the equilibrium position for a monopolist who fixes a market price subject to a known demand curve. The system has no tendency to move and is in equilibrium in the second sense but clearly demand does not equal supply and the first definition of equilibrium is inappropriate. This concept of an equilibrium which is defined by an absence of change is fundamental to much of the theoretical literature on disequilibrium or temporary equilibrium which has grown out of the work of Clower (1965) and Leijonhufvud (1968); a recent survey of this work may be found in Benassy (1982).

In its most fundamental form we can model a market with the following two equations

$$D_t = \alpha_1 P_t + B_1 Z_t + U_{1t} \tag{3.1}$$

$$S_t = \alpha_2 P_t + B_2 Z_t + U_{2t} \tag{3.2}$$

Equation 3.1 is a demand curve, which relates demand for any good to its real price (P_t), a set of other factors Z_t, which may be a vector, and U_{1t} a stochastic error term. Equation 3.2 similarly relates supply to the price of the good, a set of other factors and an error term U_{2t}. These two equations are common to all forms of market analyses; the various approaches differ in their assumption about what is observed and how the real price is determined. A full equilibrium approach would assume that $D_t = S_t = Q_t$ (where Q_t is observed); the real price is determined simply as the price which makes $D_t = S_t$. An assumption of imperfect competition often amounts to assuming that the market is dominated by a monopolist (monopsonist) and that therefore we only observe points on the demand (supply) curve, i.e. in the monopoly case $D_t = Q_t$ and the supply curve is unobservable. The distinguishing feature of the discrete disequilibrium approach is the assumption that the observed quantity actually being traded will always be on the short side of the market, that is

$$Q_t = \min(D_t, S_t) \tag{3.3}$$

The justification for this approach is based on the notion of voluntary exchange; a demand or supply curve may be thought of as defining the maximum amount of a good which will be exchanged voluntarily at a given price. If someone is offered a smaller quantity than he demands at a given price, he will generally accept this trade as profitable, but an individual will not generally purchase a larger quantity than indicated by his demand curve.

In order to close the model it is necessary to make some assumption about the determination of prices. The usual assumption which is made is that

$$P_t = P_{t-1} - \gamma(D_t - S_t) + U_{3t}, \qquad \gamma > 0 \tag{3.4}$$

so that if demand is greater than supply the real price will rise and if it is less than supply the price will fall. Equations 3.1–3.4 then constitute a full statement of the single market disequilibrium model. Over time the real price will tend to adjust to the market clearing price and the speed at which it does this is governed by γ. If γ becomes very large the disequilibrium model will very quickly move towards equilibrium. If γ is small then disequilibrium will persist for a considerable time. One of the advantages of using an empirical model based on Equations 3.1–3.4 therefore is that the estimate of γ will give us an indication of how closely the model approximates a market clearing model.

While the theoretical literature on disequilibrium models has grown very quickly over the last 15 years the literature dealing with empirical applications of the approach has remained sporadic. The reason for this lies in the difficulty of actually estimating a model of the form 3.1–3.4 where we do not know a priori whether an observation Q_t should be assigned to the demand curve or the supply curve. The paper which founded the work on empirical applications is Fair and Jaffee (1972); this paper considered a model like Equations 3.1–3.4 except that Equation 3.4, the price equation, was assumed to hold exactly. This meant that the direction of price movement could be used as an exact guide as to whether the observation was on the demand or supply curve. We can rewrite Equation 3.4, if $U_{3t} \equiv 0$ as

$$D_t - S_t = \frac{1}{\gamma}\Delta P_t$$

Now when $D_t < S_t$,

$$Q_t = S_t + (D_t - S_t) = S_t + \frac{1}{\gamma}\Delta P_t$$

So that when $D_t < S_t$ we can write 3.1 and 3.2 as

$$Q_t = \alpha_1 P_t + B_1 Z_t + U_{1t}$$

$$Q_t = \alpha_2 P_t + B_2 Z_t + \frac{1}{\gamma}\Delta P_t + U_{2t}$$

Similarly when $S_t < D_t$, $Q_t = D_t + D_t - S_t$

$$Q_t = D + \frac{1}{\gamma}\Delta P_t$$

So that we can write the model as

$$Q_t = \alpha_1 P_t + B_1 Z_t + \frac{1}{\gamma}\Delta P_t + U_{1t}$$

$$Q_t = \alpha_2 P_t + B_2 Z_t + U_{2t}$$

We can then write these two set of equations as one by defining two new variables

$$\Delta P_t^1 = \begin{cases} \Delta P_t & \text{if} \quad \Delta P_t > 0 \\ 0 & \text{otherwise} \end{cases}$$

$$\Delta P_t^2 = \begin{cases} \Delta P_t & \text{if} \quad \Delta P_t < 0 \\ 0 & \text{otherwise} \end{cases}$$

And so the following whole model can be estimated by instrumental variables (to allow for the endogeneity of P_t, ΔP_t^1 and ΔP_t^2).

$$Q_t = \alpha_1 P_t + B_1 Z_t + \frac{1}{\gamma}\Delta P_t^1 + U_{1t}$$

$$Q_t = \alpha_2 P_t + B_2 Z_t + \frac{1}{\gamma}\Delta P_t^2 + U_{2t}.$$

This solution to the switching problem is undoubtedly very elegant but it rests on two crucial assumptions: the first is that $U_{3t} = 0$ and the second that nothing other than $(D_t - S_t)$ affects the real price. Both of these assumptions are unlikely to be realistic and it is for this reason that the Fair–Jaffee approach has been used to only a limited extent. An alternative approach to the Fair–Jaffee technique was proposed by Maddala and Nelson (1974) who were able to derive the appropriate density function for a model such as Equations 3.1–3.4. This made it possible to carry out maximum likelihood estimation of such models and it is this approach which has formed the basis of most of the empirical applications and which we will use here. A survey of much of this literature may be found in Quandt (1982); we will provide a detailed account of the likelihood function derivation in Section 3.4 below. Sections 3.2 and 3.3 discuss the general issues of multimarket non-clearing models and the details of our application of the model to the UK labour market.

3.2 THE MULTIMARKET NON-CLEARING MODEL

To focus on the special features of the disequilibrium or non-clearing model, it is most convenient to start from the opposite case, that of the Walrasian equilibrium model. In this model, it is assumed that there are a number of markets and in each demands (D) and supplies (S) are dependent upon real prices (each price expressed in terms of a *numeraire* which could be the overall price level). This multimarket system of demands and supplies may, under certain conditions, be solved to provide the equilibrium set of prices and quantities. It is important to emphasize that in this familiar model it is *notional* demands and supplies which are being used. That is to say, the demand and supply functions are those which result when households and firms maximize their opportunities subject only to the constraints set by wealth and technology. Moreover, trading takes place at the 'correct' (i.e. equilibrium) set of prices, so that demands and supplies are consistent. In this scheme, it is assumed that no one makes the mistake of offering to trade at the incorrect set of prices, once trading takes place.

The non-clearing approach to the multimarket model conveys very different general implications about the determinants of demand and supply in each market and about the interrelations between markets. For our purposes the main points can be obtained by considering only two markets: the goods and labour markets. Suppose that in one market – say the goods market – trading occurs at a price above equilibrium, so that there is excess supply of goods. We assume that there are no inventories. The presence of excess supply in the goods market implies that the actual amount sold by firms is X_1 which is less than what firms plan to sell at p_1 (which is X_2). If we suppose this excess supply of goods persists, then producers' plans for employment will be affected by the actual sales of goods they obtain (X_1), not what they would like to sell (X_2). Thus the demand for labour is

$$D = f(w, X_1) \tag{3.5}$$

where w is the real wage, and is the *effective* as opposed to notional demand for labour. In effect the case outlined above treats the firm as if it were rationed in the goods market, and this gives an additional constraint (apart from technology) when the firm makes employment plans. Thus the state or regime which exists in one market exerts a *spillover* effect on the form of the demand and supply functions which describe behaviour in the other market. (It was the important contribution made by Clower (1965) which drew attention to the nature of these spillover effects and their relation to Keynesian economics.)

There are a number of additional important points to make about this analysis. Firstly, it is evident that even in the two-market case other permutations of excess demand or supply in either market are possible, and these convey alternative implications about the formulation of demand/supply functions in each market. This is an important topic, but since it will not have a practical bearing on what we will do subsequently, we will not pursue that matter here. (For extensive discussion see Malinvaud (1977).) Secondly, as our simple example illustrates, the question of the existence of a simultaneous solution to the two markets under the postulated regime still remains. Thus the demand for goods has not been specified so far. Clearly households' consumption plans will depend upon *actual* employment, in turn dependent upon the effective demand function for labour. The question is, does this multiple market actually determine a unique set of prices and quantities of goods exchanged and wages and employment levels? This question has been extensively explored under the heading of the *coherency* of regimes in a multimarket model with rationing, most notably by Gourieroux *et al.* (1980). The final question then pertains to the reason for sluggish price responses in a particular market where this market may be subject to spillover effects. This is a matter which is closely related to the model we will be describing in later sections, so we spend the rest of this section noting some of the problems which this presents.

So far we have argued that spillover effects from one market can influence the functional form of demands and supplies in another. Given these effective demands and supplies, there remains the question of the responsiveness of price in that market to excess demand or supply where these are effective demand and supplies. The labour market, for example, may be affected by the spillover of excess supply of goods. The question we pose is the responsiveness of the real wage to effective excess supply in the labour market. In the case we have used so far this is

$$\Delta w = -\gamma[f(w, X_1) - S], \qquad \gamma > 0 \tag{3.6}$$

where $f(w, X_1)$ is the effective demand for labour in this regime. At a theoretical level there may be reasons for expecting sluggish price adjustment. In the labour market, wages may be set by unions with employers setting employment levels, and real wage sluggishness could then be due to the weighting the union implicitly applies to wage increases for their members as compared to risk of unemployment and relative income out of work. A similar point applies to sluggishness in the movement of goods prices if we consider the goods market faced with, for example, excess effective supply. Again the existence of adjustment costs on price changes, or the widespread use of fixed price contracts, are reasons frequently adduced for unresponsive price behaviour (see Quandt (1982)).

In the empirical work we describe below, the presence of non-clearing in the labour market will be the principal consideration. Multiple-markets as such are not estimated. However, following our earlier discussion, spillover effects are investigated via the choice of variables which enter the demand and supply functions of labour. Then, the question of the responsiveness of the wage in the labour market is investigated by testing the proposition that the labour market itself may not clear. The techniques for determining the presence of non-clearing in a single market are then crucial to this investigation, and a detailed description of this is provided in the next section.

3.3 THE LABOUR MARKET IN DISEQUILIBRIUM

In partial analysis, the existence of disequilibrium really takes two forms: one implicit, the other explicit. The implicit part lies in the specification of the functions used, relying upon the notion of spillover effects as implied by multimarket models. Thus we could include sales of goods in the demand for labour, justifying this by assuming a sales constraint in the goods market. This, however, is implicit since it does not provide a full model of the goods market with the possibility of disequilibrium there. The explicit treatment of disequilibrium in the single market case, however, allows for the possibility that trading can take place in the market in question at prices other than the equilibrium price. Because transactions are voluntary, then quantity traded cannot exceed demand or supply. Also it cannot be less than the minimum of demand and supply, since otherwise this would mean that possibly beneficial trading opportunities were being ignored. Hence, it is assumed that quantity exchanged is determined on the 'short side' of the market. This feature is summarized in the minimum condition which is the hallmark of the single market disequilibrium case in

$$Q = \min(D, S) \tag{3.7}$$

i.e. quantity exchanged at a given price is given by whichever is the smaller, demand or supply. Figure 3.1 illustrates this.

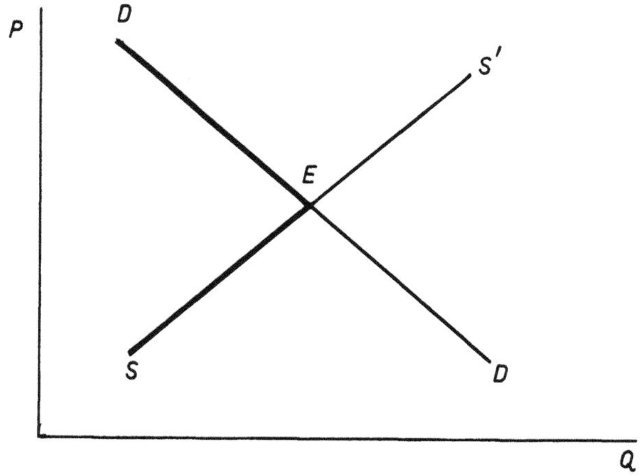

Fig. 3.1

The important property of this assumption is that, unlike the equilibrium model which implies a unique relation between price and quantity at E, the disequilibrium model implies a locus of points, given by the locus DES. In this alternative the price at which trading occurs and the amount traded could be anywhere along this locus. This property involves special problems when the model is estimated, and we will describe how these may be handled in the next section. Before that, we outline a model of the labour market which uses both the implicit and explicit features of the disequilibrium approach. We first give the detailed specification of the labour demand and supply functions, and then describe the disequilibrium labour market model.

3.3.1 The demand for labour

It is evident from what we have already said about the non-clearing model in the single market case that a range of assumptions may be made about the determinants of the underlying demand and supply functions. At one extreme orthodox neoclassical functions could be assumed, where these depend only upon the labour market. At the other, as we have emphasized, the underlying functions allow for spillover effects from the goods market. The procedure we follow will emphasize the latter interpretation, but since we use a sufficiently general econometric model of demand and supply, the resulting econometric testing will decide the appropriate specialization of the functions.

In the case of the demand function, we allow for a wide range of possible determinants of (static) demand, and, in addition, we embody a rich specification of employment dynamics by using the recent development associated with the rational expectations hypothesis (hereinafter called the REH).

To decide on the variables entering the labour demand equation, we postulate a general model. At one extreme, if the goods market is in excess demand, so there is no quantity constraint on the supply of goods, the perfectly competitive model implies an equation for employment dependent upon real prices, i.e.

$$E^{D} = f_1(W/P, P_m/P, K) \tag{3.8}$$

where W/P is the real wage, P_m/P the real price of materials, and K the capital stock. The underlying production function is assumed to be dependent upon labour, capital and materials. In all subsequent discussion we assume for simplicity that the labour/materials decision is made independently of the optimal decision by the firm regarding its capital input. Capital is then taken to be predetermined, and in the empirical applications reported below, it is represented by a time trend. It is then possible to derive an alternative equation for employment by assuming that the goods market is in excess supply so that *a sales constraint* operates on the firm, and its factor demand for labour parallels that produced by familiar cost minimizing subject to a sales constraint, so

$$E^{D} = f_2(W/P_m, Q^*) \tag{3.9}$$

where Q^* is the sales constraint. Yet another alternative, however, substitutes out the capital stock in Equation 3.8 using the production function, and this gives an equation dependent upon real prices and output (see, for example, Rosen and Quandt (1978)). The static form of the equation we use will be similar to this, and may be written in log-linear form as

$$\ln E^{\mathrm{D}} = a_1 \ln (W/P) + a_2 \ln (P_{\mathrm{m}}/P) + a_3 \ln Q^* \tag{3.10}$$

Although we do not use a model of the goods market with Equation 3.10, we none the less have to recognize that Q^* is determined jointly with employment. The interpretion we place on Q^* is that it is output ($\simeq$ sales if there is no inventory accumulation) which is expected by the firm. To make this operational, we need an output model, which we hypothesize the typical firm uses in forming its expectations about current and future sales. In our applications, we estimate a vector autogressive (VAR) model for output. This starts by assuming

$$Q_t^* = a(B) Z_t + \eta_t \tag{3.11}$$

where $a(B)$ is a polynomial in the lag operator B (hence $BX_t \equiv X_{t-1}$), and η an error term. Z_t is a set of variables thought by producers to be important in determining aggregate sales. We assume these are a measure of overall fiscal stance (BD), and a variable which is strongly influenced by monetary policy – the level of international competitiveness (C). Since, as we will show below, future values of Q^* will enter the labour demand equation, we need to set up a dynamic model for Q^*, BD and C which can be used for deriving predictions for Q^*. Such a full VAR model is

$$\begin{aligned} \ln Q_t &= a_{11} \ln Q_{t-1} + \cdots + a_{1K} \ln BD_{t-1} + \cdots + a_{1J} \ln C_{t-1} + \cdots \\ \ln BD_t &= c_{11} \ln BD_{t-1} + \cdots + c_{1N} \ln Q_{t-1} + \cdots c_{1L} \ln C_{t-1} + \cdots \\ \ln C_t &= d_{11} \ln C_{t-1} + \cdots + d_{1M} \ln Q_{t-1} + \cdots d_{1p} \ln BD_{t-1} + \cdots \end{aligned} \tag{3.12}$$

Equation 3.10 is a static formulation, but we will need to introduce dynamics since, as is well known, there are many reasons for expecting employment changes to be subject to lags. One analysis of employment dynamics which was proved successful in much applied work emphasizes lags in employment due to adjustment costs. The simplest assumption is that these costs are quadratic in the log of employment, i.e.

$$C(\Delta E) = C_1 (\Delta E)^2 + C_2 (E - E^*)^2 \tag{3.13}$$

C_1, $C_2 > 0$ where $E = \ln$ Employment. The first item in this cost function shows that there is a cost in changing employment, and that this cost increases the larger that change is. The second term penalizes changes in employment away from E^*, the static equilibrium level of employment. In turn the static equilibrium is given by a static demand equation, and in our applications we will assume that this is given by Equation 3.10.

In order to derive a full dynamic model of labour demand, we assume that the firm is subject to the adjustment cost function (3.13) above, and that it makes its employment plans optimally over time. The latter assumption can be formalized by letting the firm minimize an intertemporal cost function with respect to its employment decision.

$$\sum_{T}^{N} \delta^T C_T(\Delta E) \tag{3.14}$$

where δ is a discount factor. It may then be shown that the equation satisfying this minimization problem is

$$E_t = \lambda E_{t-1} + (1-\lambda)(1-\lambda\delta)\sum_{T} (\lambda\delta)^T E^*{}_{t+T} \tag{3.15}$$

Although we are not going to provide details on the derivation of this dynamic equation, several characteristics of it are important for subsequent sections, and we describe these now. (A full account of equations of this sort is given in Sargent (1979).) These characteristics are as follows.

The model above is *deterministic*. In this application, this means that future values of E^* are known (or perfectly predictable). This is clearly an extreme assumption. An alternative is to assume that the firm plans employment by predicting E^* using information on its determinants. This in turn means that we need a dynamic model for predicting these determinants of E^*. In our example, these are the real wage (W/P), real material prices (P_m/P), and real demand Q^* (Equation 3.10). Earlier we described a dynamic model which we will use for obtaining predictions for Q^* (Equation 3.12). For the remaining variables, the real wage and material price, there is evidence that they can be represented as random walks, i.e.

$$X_{t+1} = X_t + \mu_{t+1}$$

where X is the variable in question and μ is an error process which is serially uncorrelated. In this case predicted future values are given by the lagged value,

$$\hat{X}_{t+k} = X_t, \qquad k \geqslant 1$$

To summarize this part of our argument, Equation 3.15 above may be written in a stochastic form, as

$$E_t = \lambda E_{t-1} + (1-\lambda)(1-\lambda\delta)\sum_{T} (\lambda\delta)^j {}_{t-1}E^{*e}_{t+T} + \eta_t \tag{3.16}$$

where ${}_{t-1}E^{*e}_{t+T}$ represents the expected value of E^* for period $t+T$, using all the information up to and including period $t-1$. Notice that we also add a random disturbance to Equation 3.16 to allow for mismeasurement of variables and other non-systematic influences upon employment.

The next important feature of this model is the evident symmetry between the weights on the lagged parts of the equation and the weights on the future expected terms, given by the common parameter λ. This is a standard feature of these models (see Sargent (1979) for example, for illustrations). Our empirical results apply these symmetry conditions.

Finally, we extend the model given in Equation 3.16 to allow for an additional lag on employment. While this may appear to be a rather arbitrary change to the specification discussed so far there are two important reasons for making it. The first is that the basics of the underlying theory carry through to this enlarged case, though the technical details of the extension are not presented here. The other reason is that there are reasons for expecting a second order case, which could arise from the aggregation of two different forms of labour, manual and non-manual for example, where each is determined by an equation like Equation 3.16 above. Also many empirical studies have found second order models give a good statistical explanation of UK employment. Thus the model we will use will be of the form (logs unless otherwise stated)

$$E_t = A_0 + A_1(W/P) + A_2Q^{*e} + A_3T + A_4E_{t-1} + A_5E_{t-2} + A_6(P_m/P)_t \quad (3.17)$$

In this equation (W/P) is the product real wage, equal to gross earnings plus employee taxes, deflated by producer prices. Q^{*e} is expected future output derived according to our earlier discussion around Equation 3.12. It is important to recognize that this is a *set* of future values of Q^{*e}_{t+k}, $k = 1, \ldots M$, and for estimation purposes we have set $M = 4$, implying that firms look one year ahead in drawing up their employment plans.

3.3.2 The supply of labour

Static labour supply is taken to be

$$E^S = \beta_0 + \beta_1(W^1/P_2) + \beta_2 \mathrm{BN} + \beta_3 \mathrm{POP} + \beta_4 \mathrm{UP} + \beta_5 \mathrm{RRI} \quad (3.18)$$

W^1/P_2 is the consumption wage defined as earnings net of direct tax, deflated by consumer prices. The other variables are the real value of unemployment benefits (BN), the working population (POP), union power (UP) and the real rate of interest (RRI).

The real consumption wage (W^1/P_2) and the real interest rate (RRI) enter the equation because it is postulated that the decision by households to supply labour is an optimal one concerning the allocation of consumption and leisure over time. In this sense it is a misnomer to refer to Equation 3.18 as a static model since it in part reflects dynamic intertemporal planning. However, our use of the term dynamic is reserved for the explicit use of lags in the model, and we will consider this extension below. To return to the theory underlying Equation 3.18, it may be shown that if the representative household maximizes an intertemporal utility function depending upon consumption and employment discounted by the households rate of time preference, the resulting plan for labour supply will depend upon the real wage and the real interest rate (the nominal interest rate minus the expected inflation rate). This is the basis of the so-called 'intertemporal substitution' hypothesis of labour supply. According to this theory, an increase in the interest rate for example has a similar effect on the allocation of leisure (the inverse of work effort) as it does upon the allocation of consumption. That is, in the case of labour supply it produces a substitution effect in favour of future leisure, or equivalently it stimulates an increase in current work effort relative to work effort in a future period. (For a full account of this hypothesis, see Barro (1984).)

The other important variables in Equation 3.18 are the level of real benefits and the variable representing union power (UP). There are a number of reasons for expecting an influence of real benefits upon labour supply. Perhaps the simplest approach uses the 'replacement ratio' (the ratio of net income when in work compared to that when unemployed). According to this, the decision either to enter or withdraw from the workforce depends upon the relative price of working compared with not working, as measured by the replacement ratio. Increases in unemployment benefits *ceteris paribus* will encourage the marginal worker to withdraw from the employed labour force. Notice that in Equation 3.18 if it were the replacement ratio which influenced labour supply decisions, we would expect equal and opposite signs on the real wage and the level of real benefits. Thus our specification of the labour supply equation allows for this possibility. The model also allows for an alternative interpretation, namely that it is the level of real benefits *per se* which influence labour supply. Minford *et al.* (1985) present this argument. On this interpretation, the average level of real benefits sets a floor to the determination of aggregate wages in the non-unionized sector of the economy. Changes in the level of real unemployment benefits then change the supply of labour in the non-unionized sector in the same direction. In keeping with our earlier approach to the demand side, we have chosen a general specification for supply which enables each of these alternative influences of unemployment benefits to emerge as a special case. Which is the most appropriate may then be judged according to the evidence. Finally, in the determinants of the static supply equation we have allowed for an effect from union power. This is done in recognition of the emphasis now being placed on explicit bargaining models of the labour market. (Good examples are Nickell and Andrews (1983) and Minford *et al.* (1985).) These emphasize the role of the union in setting real wages according to a union utility function dependent upon expected real income of its potential membership, given the firm's demand curve. Many alternative bargaining structures are possible, including ones where the union sets the real wage unilaterally and the firm sets employment unilaterally. It is not possible, or indeed suitable, that we review these here. Rather our approach is to note that these models attribute a role for union power (somehow defined) in establishing or helping to establish the real wage. In our applications we allow for this influence by a direct effect upon the supply of labour, e.g. if an increase in union power increases the real wage at given employment levels, this is equivalent to a leftwards shift in the supply of labour in the unionized sector.

The remaining question is that of dynamics in the labour supply equation. We have mentioned already that the model given by Equation 3.18 is implicitly dynamic as it is based, in part, on the idea that households maximize an intertemporal utility function. There are additional reasons for expecting explicit dynamic effects, including lagged labour supply responses to changes in variables on the right-hand side of Equation 3.18. However, unlike the case with the demand equation, it is not feasible to base the complete model – both static and dynamic parts – upon the solution of one intertemporal maximizing problem. As will be obvious already, the labour supply model we are proposing, synthesizes a number of distinct approaches, the aim being to assess the empirical role of these alternatives. Thus we allow for the possibility that there may be lagged adjustment in labour supply, and extend Equation 3.18 to incorporate two lags in employment i.e.

$$E_t^S = \beta_0 + \beta_1(W^1/P_2)_t + \cdots + \beta_6 E_{t-1} + \beta_7 E_{t-2} \tag{3.18}$$

The rationalization for these lags is that there are costs of adjustment on the labour supply side, due to for example losses incurred due to imperfections in the capital market, or due to loss of seniority rights (see Andrews (1983)). Also, although the intertemporal nature of the supply equation suggests that households use forward expected values of, for example, real wages and real interest rates, we will not experiment with such versions of the model here. The reason being that we assume forward-looking households use the current values of these real variables since these convey sufficient information about their expected future levels. This parallels the assumption about the expected real product wage which was made in Section 3.3.1 above. Formally, it entails that the real variable in question behave as a 'random walk', so that the change in the variable is a random uncorrelated series.

Having outlined the reasoning behind the labour demand and supply equations we will be using, we will next describe some general problems in estimating disequilibrium models. This description is then followed by the empirical results which we obtain in estimating the model we have described.

3.4 EMPIRICAL RESULTS

In this section we first describe the methods we use to estimate the labour market when it is assumed to be in disequilibrium. Then we will describe results of estimating the model outlined in the previous section.

3.4.1 Estimating disequilibrium models

In its simplest form, the disequilibrium model may be written as a set of linear equations

$$\begin{aligned} E^D &= X_1'\beta_1 + \alpha_1 P + u_1 \\ E^S &= X_2'\beta_2 + \alpha_2 P + u_2 \\ Q &= \min(E^D, E^S) \\ \Delta P &= \gamma_1[E^D - E^S] + \gamma_2 Z + u_3 \end{aligned} \tag{3.19}$$

where X_1 and X_2 are determinants of demand and supply respectively, u_i (1, 2, 3) are random disturbances. In this model, variables other than excess demand are assumed to produce changes in P (for the labour market P would be the real wage, of course) and the presence of these is indicated by Z in the last equation. To derive the likelihood function, let

$$f(u_1, u_2, u_3) \tag{3.20}$$

be the joint distribution of error terms, and $g(E^D, E^S, P)$ the joint density of E^D. E^S and P derived from $f(.)$. *Observations* (Q) on the *demand* side are

$$E^D = Q, E^S > Q$$

with distribution for Q given by

$$h(Q, P|Q = E^D) = \left[\int_Q^\infty g(Q, E^S, P)\, dE^S\right](1/Pr(Q = E^D))$$

similarly for *supply*

$$h(Q, P|Q = E^S) = \left[\int_Q^\infty g(Q, E^D, P)\, dE^D\right](1/Pr(Q = E^S))$$

Hence, we may write $h(Q, P)$ as

$$h(Q, P) = \int_Q^\infty g(Q, E^S, P)\, dE^S + \int_Q^\infty g(Q, E^D, P)\, dE^D$$

In the case we consider, the joint densities of the u's are assumed to be normally distributed with zero mean and diagonal covariance matrix

$$V = \begin{vmatrix} \sigma_1^2 & 0 & 0 \\ 0 & \sigma_2^2 & 0 \\ 0 & 0 & \sigma_3^2 \end{vmatrix}$$

The density $g(.)$ may be written

$$g(.) = \frac{H(t)\exp\left(-\tfrac{1}{2}U^t V U\right)}{(2\Pi)^{3/2}\sigma_1\sigma_2\sigma_3}$$

where $U' = (u_1, u_2, u_3)$ and $H(t) = 1 + \gamma_1(\alpha_2 - \alpha_1)$

Assuming the observations are independent, the likelihood function is

$$L = \sum_t^T \ln h(Q_t, P_t)$$

where T is the sample size.

3.4.2 Empirical results

For convenience, we restate the general model again. It is of the following form (unless otherwise stated variables are in logs throughout):

$$E^D = A_0 + A_1(W/P)' + A_2Q^e + A_3T + A_4E_{-1} + A_5E_{-2} + A_6\text{RP}$$
$$E^S = B_0 + B_1(W/P) + B_2\text{BN} + B_3\text{POP} + B_4\text{UP} + B_5\text{RRI} + B_6E_{-1} + B_7E_{-2}$$
$$E = \min(E^D, E^S)$$
$$(W/P)' = D_0(W/P)$$
$$\Delta_4(W/P) = C_0 + C_1(E^D - E^S) + C_2\Delta_4\text{UP} + C_3\Delta_4\text{NTAX} + C_4\text{IP} \quad (3.21)$$

(IP, UP and NTAX are not in logs, $\Delta_4 X_t \equiv X_t - X_{t-4}$)

In these equations $(W/P)'$ is the demand price of labour, given by gross earnings per employee plus employer taxes deflated by producer prices. The variable Q^e is expected output (= expected sales) described earlier. Real raw materials and energy prices in real terms are given by RP. Finally, the remaining variable in the demand equation is time (T). The supply function includes the supply price of labour (the *net* level of earnings per worker deflated by the price of consumer goods; the real level of unemployment benefit (BN), the working population (POP), a measure of union strength (UP) and the real interest rate (RRI). In the last equation, NTAX is employers' tax as a proportion of employees' income, and IP is an incomes policy variable reported by Whitley. (This is described in Whitley (1983).)

The key equations in the model have already been described at some length.

Table 3.1 Disequilibrium estimates of the UK labour market
Sample: Quarterly 1964(4)–1982(4)[1] for Equation 3.1
Quarterly 1964(4)–1980(1) for Equation 3.2

Parameter	*(1)*		*(2)*	
A_0	0.072	(0.5)	−0.44	(1.1)
A_1	−0.014	(2.9)	−0.018	(3.2)
A_2	0.042	(5.8)	0.068	(3.7)
A_4	1.588	(119.0)	1.637	(308.0)
A_5	−0.630	(21.0)	−0.510	(7.4)
A_6	−0.007	(1.6)	−0.013	(2.3)
B_0	−0.287	(1.9)	−0.187	(0.8)
B_1	0.083	(4.0)	0.092	(4.9)
B_2	−0.041	(4.8)	−0.048	(8.8)
B_3	1.179	(26.0)	1.153	(18.8)
B_4	−0.188	(14.5)	−0.187	(17.9)
C_0	0.029	(6.9)	0.028	(5.8)
C_1	0.223	(2.21)	1.124	(3.5)
C_3	−0.651	(3.6)	−0.556	(2.7)
C_4	0.007	(5.7)	0.005	(4.2)
log likelihood function	626.68		559.459	

[1] Unlike other chapters in this book we have used quarterly data for our estimation. It is very important to use the largest possible data set for the maximum likelihood estimation of the sort used here, and this precludes the use of an annual data set.

It remains to note that the fourth equation linking the two definitions of the real wage ($(W/P)'$ the product real wage and (W/P) the consumption real wage) is a technical relation, enabling the model to be expressed in terms of two variables (W/P) and E, so that the model may be estimated using the techniques described in Section 3.4.1 above. The final equation is the wage adjustment equation, where the adjustment is fourth quarter changes. Apart from being determined by excess demand for labour, this equation also allows for autonomous effects from other variables. Thus a proxy for union power (UP), employers' contributions as a proportion of employees' income (NTAX) and an index of incomes policy (IP) are used in the general form of adjustment equation, to establish whether they have any explanatory power. In the case of UP and IP, many studies have suggested that each effect wage inflation. In the case of the NTAX variable, this is included to establish whether wage bargains are made in terms of the supply real wage or the demand real wage (if $C_3 = -1$ it is the demand real wage, if $C_3 = 0$ it is the supply real wage).

In Table 3.1 we display the empirical results for the disequilibrium model. This is our preferred form of the model and has been arrived at by experimentation with the general model, dropping insignificant or implausible variables. We describe the most important of these now. A more extensive discussion of this search for the preferred model with the empirical results for a number of alternatives is provided by Hall *et al.* (1986).

In brief, our simplification search considered the results for the model using the complete sample (column 1). What this revealed was that it was not possible to obtain well-determined lagged effects in the supply equation (B_6 and B_7 are the relevant parameters), and so our preferred model dropped these variables. In the demand equation, we were able to identify correctly signed real wage and real demand effects (parameters A_1 and A_2 respectively), but the time trend which proxies trend labour productivity and disembodied technical progress proved to be insignificant. In the real wage adjustment equation, our proxy for union power (UP) was also insignificant and so was excluded from that equation. The same variable regularly had a highly significant effect upon labour supply however (parameter B_4), so the model as a whole does incorporate significant effects from union power. The remaining variables in the model are then generally well determined with plausible signs according to a priori conventions. The effect of real raw material prices is not significant at the conventional 5% level (A_6), but it was thought desirable to keep this variable in the preferred model. This apart, the model is highly satisfactory from a statistical point of view.

To interpret these results, two crucial distinguishing features of this model have to be emphasized. The first is that the model is composed of a demand, supply and wage adjustment equation. In this it encompasses dynamic equilibrium or new classical models familiar from the Rational Expectations literature. Our results show that there is strong statistical support for including a wage adjustment equation. An equilibrium model, for example, would imply that excess demand terms in the real wage adjustment equation should be infinite. Our results reject this. What needs to be emphasized about this finding is that the behavioural equations for labour demand and supply we use are based on intertemporal optimizing theory familiar from the Equilibrium School. Hence, although we find considerable empirical support for many parts of these specifications, we also find a significant role for disequilibrium real wage adjustment.

The other feature of our results is even more distinctive, and concerns the explicit inclusion of the 'min' condition, or short-side constraints in the model. Given the estimated parameters, it is then possible to apportion employment as being determined either by the supply or demand side of the market. Figure 3.2 shows the *ex post* allocation of employment between demand and supply constraints. What this indicates is that for much of the period, demand supply and actual employment are fairly close to each other. This suggests that the model implies something close to a stochastic equilibrium in the labour market over much of the sample. However, when we consider the period 1980–1982, the implications are startlingly altered. Here the massive fall in aggregate employment is attributed entirely to a fall in the aggregate demand for labour. The last part of this period therefore is strongly characterized in this work as a period of substantial excess supply.

Pursuing this point further, we re-estimate the model dropping the last eight observations (1980(1)–1982(4)). The results of this are shown in column (2) of Table 3.1. In the main, the estimated results are very similar to those obtained in the full sample, and suggest that the main features of the model are robust to

Fig. 3.2 Comparison of labour demand and supply

sample changes. However, the adjustment parameter in the real wage equation (C_1) changes dramatically in the shorter sample, being very substantially larger. According to this result the responsiveness of real wages to excess demand/supply is much *reduced* over the long sample, a period substantially affected by the events of the last two years of data. This suggests that labour market sluggishness – as indicated by the response of real wages to changes in excess demand or supply – has worsened as compared with its apparent behaviour up to 1980.

3.5 POLICY IMPLICATIONS

The model set out in Equation 3.21 is, in many ways, an unremarkable one. Demand is a function of output and relative prices, supply is again a function of relative prices, population, union power, benefits and interest rates. These general functional forms may be found in many studies of the labour market (e.g. Minford *et al.* (1985), Nickell and Andrews (1983)). The policy implications of the model estimated above are however fundamentally different from those flowing from the more conventional market clearing or bargaining models. The most important difference comes from the switching nature of the model. At any moment in time our model views the labour market as being either demand constrained or supply constrained. The effect of changing one of the inputs to the labour market (e.g. benefits, union power, or expected output) will depend crucially on the regime which is dominating the labour market at that moment.

This may be illustrated in Figs 3.3 and 3.4. These two diagrams illustrate the effect of implementing a policy aimed at increasing the supply of labour; this might be in the form of a reduction in unemployment benefits or a reduction in the union power variable (ignoring for present purposes how that might be achieved). For simplicity in these two diagrams we have assumed that the real wage is fixed; we will relax this assumption in a moment. In Fig. 3.3 employment

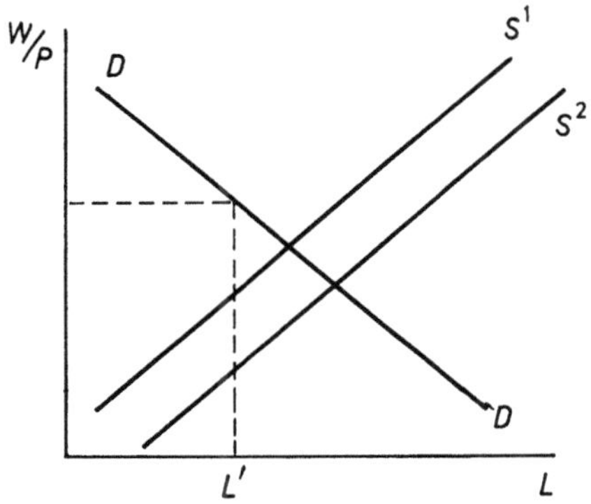

Fig. 3.3

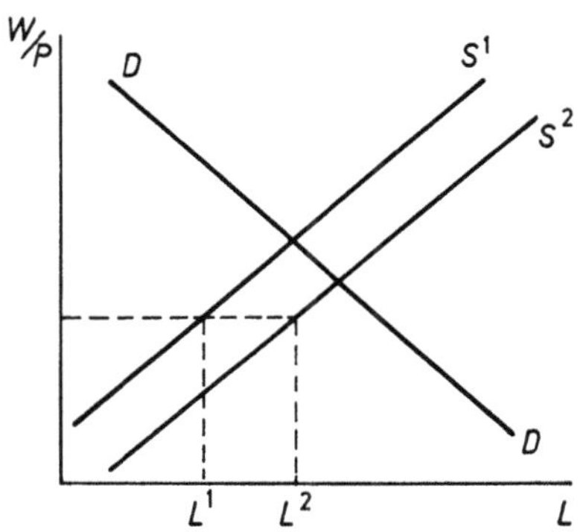

Fig. 3.4

is determined on the demand side of the market, so a policy aimed at the supply curve has no direct effect on employment at all. In Fig. 3.4 the labour market is dominated by the supply function and so a shift in the supply curve has both large and powerful direct effects on employment. A similar illustration could have been given for a change affecting the demand curve; a rise in expected output for example would be a powerful policy when the labour market is demand constrained but will be ineffective when the market is supply constrained.

This simple illustration does not contain the full story of course; wages are held fixed whereas in our model the wage would respond to any change in demand or supply, as the dynamic equation for the real wage in Equation 3.21 depends upon actual excess demand of labour ($E^D - E^S$). Furthermore, as the empirical results in Table 3.1 show, there are lags in the demand equation, and these imply that the full response of demand to changes elsewhere in the model take time to build up. In order to calculate the response of the model to changes in an exogenous variable, we must therefore solve the complete model for each time period. This solution will then allow for feedback between demand, supply and real wages, and for all dynamic adjustments to take place. To investigate complete solutions to the model we have simulated the effects on the preferred model in Table 3.1 of changes in two quite different variables which affect supply and demand. The two we have selected are a supply-side variable, the level of real unemployment benefits, and a demand-side variable, the level of expected output. To illustrate our argument that the discrete switching model we are advancing has very different policy responses, compared with traditional models, we have computed these simulations in each case according to either of two dominant regimes: a demand or a supply constrained regime. That is to say, we investigate the effects of, for example, a 10% reduction in real unemployment benefits, assuming first that the market is demand constrained throughout the period. Next we calculate the effects of the same change in real unemployment benefits, now assuming the market is supply constrained throughout. The results

Table 3.2 The effect of reducing benefits by 10% (percentage change in employment in the 4th quarter of the year)

Year	*Demand constrained*	*Supply constrained*
1	0.01	0.4
2	0.03	0.4
3	0.06	0.4
4	0.09	0.4
5	0.12	0.4
6	0.14	0.4
7	0.17	0.4
8	0.19	0.4
9	0.20	0.4
10	0.23	0.4

Table 3.3 The effect of increasing expected output by 10% (percentage change in employment in the 4th quarter of the year)

Year	*Demand constrained*	*Supply constrained*
1	2.8	0.0
2	5.6	0.0
3	7.5	0.0
4	7.5	0.0
5	8.5	0.0
6	7.5	0.0
7	7.5	0.0
8	6.6	0.0
9	6.6	0.1
10	5.6	0.1

of these two simulations are shown in Table 3.2. In Table 3.3, we report on the simulated effects of a 10% increase in expected output, done on the same basis.

Table 3.2 provides quantified information on the sort of phenomenon suggested by Figs 3.2 and 3.3. In the second column the change in employment is given by the estimated coefficient in the labour supply function B_2. This is estimated to be -0.041 (see Table 3.1). The decrease in real unemployment benefits thus produces a quick full effect, as anticipated, given the supply constraint. In column 1, with a demand constraint in operation, the quantitative effects are quite different. Not only do they start slowly, but even after ten years their effect is only a little more than half that of the previous exercise. The *modus operandi* in the demand-constrained case is as follows. The reduction in real unemployment benefits shifts the supply curve to the right lowering excess demand for labour, and hence slows down the pace of real wage change. This will tend to increase demand – compared with what it would have been given unchanged real unemployment benefits — via the real wage effect in the demand equation. What our results indicate is that this is not a particularly powerful effect, however, and, moreover, it takes a considerable time to build up.

Turning to our second example, we now simulate the model by changing the expected level of output. The results of this for each regime are shown in Table 3.3.

What this result shows is that in the supply constrained regime, the change in expected output is almost totally ineffective in changing employment. In the demand constrained regime, however, it has a large and quite quick acting effect. After five years the resulting change to employment virtually reaches the value implied by the long-run elasticity on real output ($A_2/(1 - A_4 - A_5)$ in Equation 3.21). Thereafter the induced increases in the real wage, due to increases in the excess demand for labour, tend to lower employment. Nonetheless, the effect after ten years is still a considerable one.

The lesson from these two exercises is a clear one: the correct policy to encourage employment depends crucially upon the state of the labour market. Supply-side policies are most effective when the supply constraint is determining employment, and similarly demand policies are most effective when the regime is one of demand constraint. Although this is an evident property of the discrete switching model we have developed in this chapter, it must be emphasized that the model we have used includes others as special cases. In particular the hypothesis that the labour market is in continuous equilibrium, as suggested by new classical macroeconomists, is a special case of our model (Equation 3.21). We have used general dynamic models for the underlying demand and supply functions which include models such as those proposed by Sargent (1979) and Barro (1984), and our estimation technique allows the data to decide for or against the presence of switches in regime. It decisively finds against equilibrium, especially so in the period post-1979, when there is strong evidence of a powerful demand constraint producing large falls in employment over this period. Hence, although our model does include factors like real unemployment benefits and proxies for union strength as influences upon supply, it clearly indicates that demand policies are the only significant way to produce substantial changes in employment.

3.6 CONCLUSIONS

Beginning from a model of the labour market which is heavily dependent upon new classical equilibrium models, we have described an extension which allows for systematic non-clearing in the labour market. Our results indicate a considerable degree of empirical support for this extended model, and this suggests that meaningful estimates of a disequilibrium model can be achieved, in spite of suggestions which are often made to the contrary. Specifically our results show that the sharp falls in employment after 1979 were due to a rapid fall in labour demand. Moreover, if anything, the labour market appears characterized by a greater degree of inertia in this latter period, as compared with its average behaviour before 1979. The policy implication of this model would strongly support the use of conventional demand management techniques to remedy this situation.

REFERENCES

Andrews, M. (1983) The aggregate labour market – an empirical investigation into market clearing. Centre for Labour Economics, London School of Economics, discussion paper 154.

Barro, R. (1984) *Macro Economics*, John Wiley.

Benassy, J. P. (1982) *The Economics of market disequilibrium*, Academic Press, New York.

Clower, R. W. (1965) The Keynesian counter-revolution: a theoretical appraisal, in *The Theory of Interest Rates* (eds F. M. Hahn and F. P. R. Brechling) Macmillan, London.

Fair, R. C. and Jaffee, D. (1972) Methods for Estimation for Markets in disequilibrium, *Econometrica*, **40**, 497–514.

Gourieroux, C., Laffont, J. and Monfort, A. (1980) Disequilibrium econometrics in simultaneous equations systems, *Econometrica*, **48**, 75–96.

Hall, S. G., Henry, S. G. B., Markandya, A. and Pemberton, M. (1986) The UK labour market: expectations and disequilibrium, National Institute of Economic and Social Research, London (mimeo).

Leijonhufvud, A. (1968) *On Keynesian Economics and the Economics of Keynes, a study in monetary theory*, Oxford University Press, Oxford.

Maddala, G. S. and Nelson, F. D. (1974) Maximum likelihood methods for models of markets in disequilibrium, *Econometrica*, **42**, 1013–30.

Malinvaud, E. (1977) *The Theory of Unemployment Reconsidered*, Basil Blackwell, Oxford.

Minford, A. P. L., Ashton, P., Peel, M., Davies, D. and Sprague, A. (1985) *Unemployment:* Cause and Cure, 2nd edn, Basil Blackwell, Oxford.

Nickell, S. and Andrews, M. (1983) Unions, real wages and employment in Britain, 1951–1979, *Oxford Economic Papers*, 35, 183–206.

Quandt, R. E. (1982) Econometric disequilibrium model, *Econometric Reviews*, **1**, No. 1.

Rosen, H. S. and Quandt, R. E. (1978) Estimation of a disequilibrium aggregate labour market, *The Review of Economics and Statistics*, **60**, No. 3, 341–9.

Sargent, T. (1979) *Macroeconomic Theory*, Academic Press, New York.

Whitley, J. (1983) Endogenising incomes policy, Institute for Employment Research, University of Warwick, Coventry, discussion paper 23.

4

A neoclassical model of the UK labour market

MICHAEL BEENSTOCK
and
PETER WARBURTON

4.1 INTRODUCTION

In this chapter we try to show how the UK labour market might be modelled along lines that are broadly neoclassical. The specification of the demand side of the labour market has much in common with the treatment of the demand for labour in other chapters. However, we experiment with labour demand schedules that assume either perfect or imperfect competition. On the other hand, unlike Chapter 3, we do not experiment with Keynesian specifications in which costs are minimized with respect to a constraint in terms of expected effective demand. Indeed this is one of the senses in which our model is neoclassical.

Wages are hypothesized in terms of a Phillips curve in which the 'natural' or equilibrium rate of unemployment is a variable. This specification implies that in the short run the labour market does not clear, however, but it tends to clear in the long run. In this respect our efforts overlap with those reported in Chapter 3 where wage growth reflects the pressure of excess demand in the labour market. However, they should be contrasted with the union-bargaining models that are reported in Chapters 5 and 6. Indeed, our efforts take no direct account of the existence of trade unions. This partly reflects our inability to generate reliable estimates of union power. In short, we think that union power matters, but we are at a loss how to measure it. The tendency for the labour market to clear is neoclassical rather than new classical. The latter implies that the labour market is always in equilibrium which is not the case in our model.

Our most innovative efforts pertain, perhaps, to our treatment of labour supply. This has several features. First we try to explain the working population in terms of male and female participation behaviour. In the case of males we take account of the fact that retirement decisions will influence the male participation ratio while in the case of females we examine the roles of equal-pay legislation, children and wages. We also investigate whether female participation is influenced by 'discouraged worker' effects according to which women do not register when unemployment is high.

We think these effects are important because a complete account of unemployment must include the behaviour of the working population as well as the number of employees in employment. This contrasts with standard treatments that do not pay serious attention to the determinants of the working population.

But the effective supply of labour is not equal to the working population because the 'natural' rate of unemployment is non-zero. We suggest that the 'natural rate hypothesis' may be used to determine the proportion of the working population that will make their labour supply effective. Our specification of the supply-side of the labour market is also broadly cast along neoclassical lines.

Our model is estimated from annual data over the period 1950–1985 and it explains the following variables: business sector employment; unemployment; earnings; male working population; female working population. General government employment is assumed to be exogenous while we follow Zabalza and Tzannatos (1985) in assuming that the difference between male and female earnings reflects equal pay legislation and is therefore exogenous. Our model does not attempt to explain endogenous variables such as inflation, GDP and the exchange rate. However, Beenstock *et al.* (1986) describe how the labour market model that is proposed here can be integrated into a broader model of the UK economy as a whole.

In Section 4.2 we set out our theoretical premises. Section 4.3 reports our principal econometric findings. Finally, in Section 4.4 we carry out a number of simulation exercises on the model that we have estimated. We also draw attention to the policy implications of our model.

4.2 THEORY

4.2.1 Labour demand

In our model the demand for labour has two elements, the demand for labour by general government (central government and local government) and the demand for labour by the business sector. The former is assumed to be exogenous since in principle it reflects public expenditure policy. The latter is endogenous and is one of the main objectives of our econometric analysis. For information we plot the behaviour of the two components of the demand for labour in Fig. 4.1. Thus:

$$L^{D} = L_{g} + L \tag{4.1}$$

where L^{D} is total demand for labour, L_{g} is general government employment and L is business sector employment.

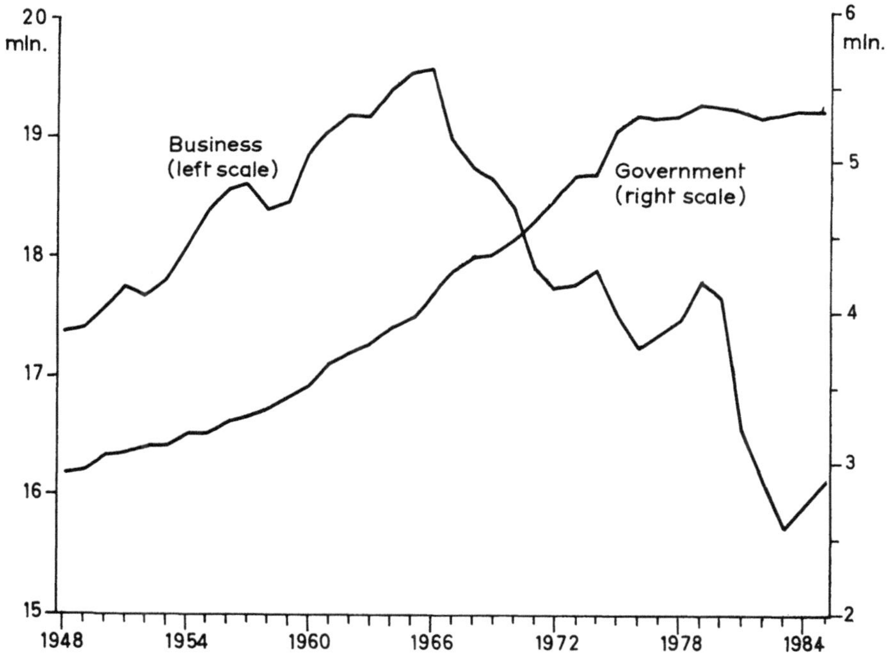

Fig. 4.1 Business and government employment

To begin with, we assume that the business sector operates under conditions of perfect competition and that firms maximize profits. Their gross output production function is characterized by Equation 4.2

$$Q = F(K, L, M, E) \tag{4.2}$$

where Q is gross output, K is capital stock, M is raw material and E is energy consumption.

The first two factors of production in Equation 4.2 are standard. The second two are singled out because over the postwar period there have been substantial changes to their factor prices which in theory and in practice affected the demand for labour. Notice that we do not hypothesize the production function in separable form as

$$Q = F(V(K, L), E, M)$$

where V denotes net output. In this case separability implies that capital and labour combine to produce a package of productive services which in turn combine with energy and raw materials to produce gross output. It also implies that the capital–labour ratio is decided independently of other factor utilization ratios. Although nested production functions of this type are frequently specified, e.g. Bruno and Sachs (1985), we cannot think of any theoretical justifications for nesting and so retain the non-nested form of Equation 4.2. As will be seen in

Section 4.3 it turns out that the capital–labour ratio is indeed dependent upon energy and raw materials factor prices.

Profits are defined as

$$\Pi = QP - W(1+e)L - KC - MP_{\mathrm{m}} - EP_{\mathrm{e}} \tag{4.3}$$

where P is the gross output price at factor cost, W is the wage rate, e is the rate of employer national insurance contributions and superannuation, C is the user cost of capital, P_{m} is the price of raw materials and P_{e} is the price of industrial energy.

If competition is perfect P is a given for each firm. Substituting Equation 4.2 into 4.3 and maximizing the result with respect to the four factors of production generates a set of integrated factor demand relationships:

$$\begin{aligned} L &= L(W(1 \mp e)/P, C/P, P_{\mathrm{e}}/P, P_{\mathrm{m}}/P) \\ K &= K(W(1+e)/P, C^{-}/P, P_{\mathrm{e}}/P, P_{\mathrm{m}}/P) \\ M &= M(W(1+e)/P, C/P, P_{\mathrm{e}}/P, P_{\mathrm{m}}^{-}/P) \\ E &= E(W(1+e)/P, C/P, P_{\mathrm{e}}^{-}/P, P_{\mathrm{m}}/P) \end{aligned} \tag{4.4}$$

The signs of partial derivatives are indicated over the variables to which they refer. In Equation 4.4 only the own factor price terms can be signed unambiguously. Competing factor price terms are ambiguous as illustrated in Fig. 4.2. For simplicity we assume only two factors of production L and K. In the initial situation the factor price ratio $W(1+e)/C$ is represented by FF′. If output remains unchanged the demand for labour must rise since producers will re-optimize at a point like b. But the profit maximizing level of output will

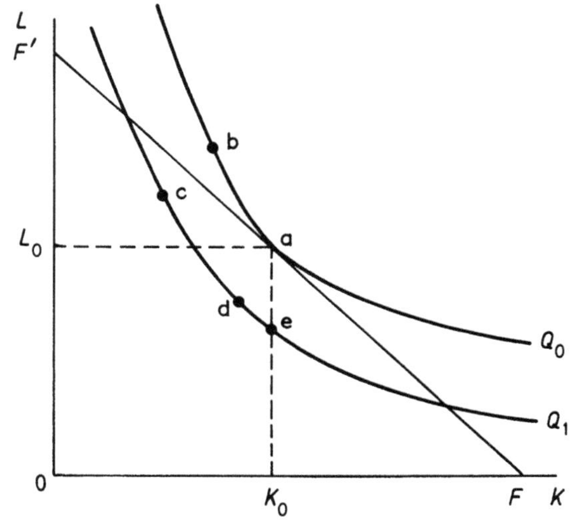

Fig. 4.2 Competing factor price ambiguity

contract (provided there are diminishing returns to scale) and production will fall to Q_1.

If the new factor price ratio is tangent to Q_1 at c the demand for labour will rise. However, there is nothing to prevent the tangent from being at a point like d in which case the demand for labour varies inversely with C. The only restriction is that the new equilibrium cannot be to the right of e along Q_1 for otherwise we would be violating the restriction that own factor price effects are unambiguously negative so that K_1 must be less than K_0.

If the solution is at c labour is a gross substitute with respect to capital. If the solution is at d it is a gross complement since falls in K induce falls in L. In Equations 4.4 we do not know whether or not factors of production are gross complements or gross substitutes with respect to each other. This can only be decided empirically as described in Section 4.4. The labour demand function in Equations 4.4 may be written as

$$L = L(\overset{+}{K}; W(1 \mp e)/P, P_e/P, P_m/P) \qquad (4.5)$$

i.e. the capital stock is assumed to be predetermined when profits are maximized. This is the approach that is applied in Chapter 6. Alternatively, we may regard K in Equation 4.5 as an endogenous variable that is determined simultaneously with other factor demands as Fquations 4.4 imply. In this more plausible case we should treat K as an endogenous regressor for purposes of econometric estimation. If returns to scale are constant it is easy to show that L is strictly proportionate with respect to K in Equation 4.5. If the elasticity of L with respect to K exceeds unity returns to scale must be increasing.

Equation 4.5 is our basic labour demand schedule by the business sector when competition is assumed to be perfect. If, instead, competition is imperfect each producer cannot take the price as given; he has to make prices rather than take them because the demand schedule for his product slopes downwards. In Chapter 6, Nickell shows that under imperfect competition Equation 4.5 still obtains at the aggregate level provided the price elasticity of demand for products is constant. If, instead, these elasticities depend on the level of demand, the aggregate demand for labour will also depend on the level of aggregate demand (D), i.e. we have

$$L = L(K; W(1 + e)/P, P_e/P, P_m/P; D) \qquad (4.6)$$

If the elasticity of demand falls in a boom the partial derivative of L with respect to D will be positive because entrepreneurs will be able to raise profit margins which will make employment more profitable. If, however, the elasticity is constant, Equations 4.6 and 4.5 become observationally equivalent.

4.2.2 Labour supply

The working population consists of males and females:

$$\bar{L} = L_m + L_f \qquad (4.7)$$

Fig. 4.3 Male and female participation ratios

where $\bar{L}$ is working population, L_m is male workers and L_f is female workers. Figure 4.3 plots male and female participation ratios. Male participation has tended to decline while female participation has tended to rise. It turns out that in the former case the decline almost entirely reflects male early retirement. For males under the age of 50, participation has been close to 100% of the able-bodied population. These observations suggest the following scheme:

$$\begin{aligned} L_m &= L_m > 50 + L_m < 50 \\ L_m < 50 &= \text{POP}_m < 50 \\ L_m > 50 &= (\text{POP}_m > 50) F_1(\text{PEN}/P_c, W(1-t)/P_c) \end{aligned} \tag{4.8}$$

where $L_m < 50$ = male workers less than 50 years, $L_m > 50$ = male workers aged 50 years and over, $\text{POP}_m < 50$ = male population of working age less than 50 years, $\text{POP}_m > 50$ = male population aged 50 years to 70 years, PEN = pension, P_c is the consumer price index and t is the standard rate of tax plus employee national insurance contribution rate.

Equations 4.8 imply that older males' retirement decision depends on the real value of pension benefits and the real disposable wage. The more generous the pension in either absolute or relative terms the less likely it is that older men will participate in the labour market. If the retirement decision only depends upon substitution effects $\text{PEN}/W(1-t)$ enters $F_1(\quad)$ in ratio form on its own. However, if income effects influence retirement decisions, the form of Equation 4.8 is appropriate.

In the case of females our specification draws on the effort of de Cooman *et al.* (1985). They suggest that apart from depending upon disposable real wages that females earn, female participation is likely to vary inversely with child-bearing. Young children, especially those under 5 years, are most likely to require the mother to remain at home. In addition, female participation is likely to reflect the state of the labour market. Women are more likely to register when job prospects are more favourable than when they are less favourable. In Equation 4.9 we assume that the latter vary inversely with the perceived rate of unemployment during a given time period.

These circumstances suggest the following hypothesis about female participation

$$L_f = \text{POP}_f F_2(\theta W(\overset{+}{1} - t)/P_c, \bar{\text{C}}\text{H}, \bar{\bar{\tilde{U}}}) \tag{4.9}$$

where POP_f is the female population of working wage, θ is female wage rates as a fraction of W, CH is the ratio of children aged 5 years or less to the female population aged 20–34 years (dependency ratio) and $\tilde{U}$ is the expected unemployment rate.

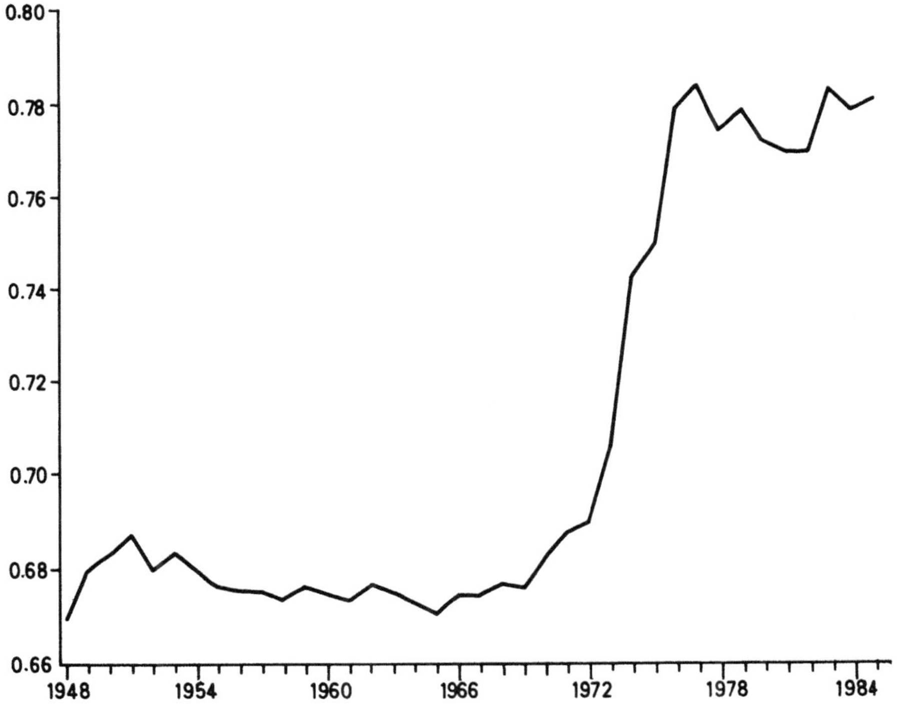

Fig. 4.4 Ratio of female to average earnings

For information we plot θ and CH in Figs 4.4 and 4.5. Relative female pay rose sharply during the 1970s.

Equations 4.7–4.9 determine the total working population (excluding the self-employed). But the effective supply of labour (L^s) does not equal L. This is because the 'natural' rate of unemployment (U^*) is not zero. The effective supply of labour is defined as:

$$L^s = (1 - U^*)\overset{-}{L} \tag{4.10}$$

i.e. we deduct voluntary unemployment from the working population in order to measure the number of people who genuinely wish to supply their labour. Because U^* cannot be directly observed, nor can L^s be directly observed. Nevertheless, we hypothesize that the 'natural' rate of unemployment varies directly with the replacement ratio and the absolute value of real social security benefits (B), i.e.

$$U^* = U^*(\overset{-}{B/W}, \overset{-}{B}/P_c, \overset{+}{SC}) \tag{4.11}$$

The former reflects the substitution effect between job market search and employment. The latter reflects the income effect whereby higher absolute benefits make voluntary unemployment more attractive as workers trade off leisure against

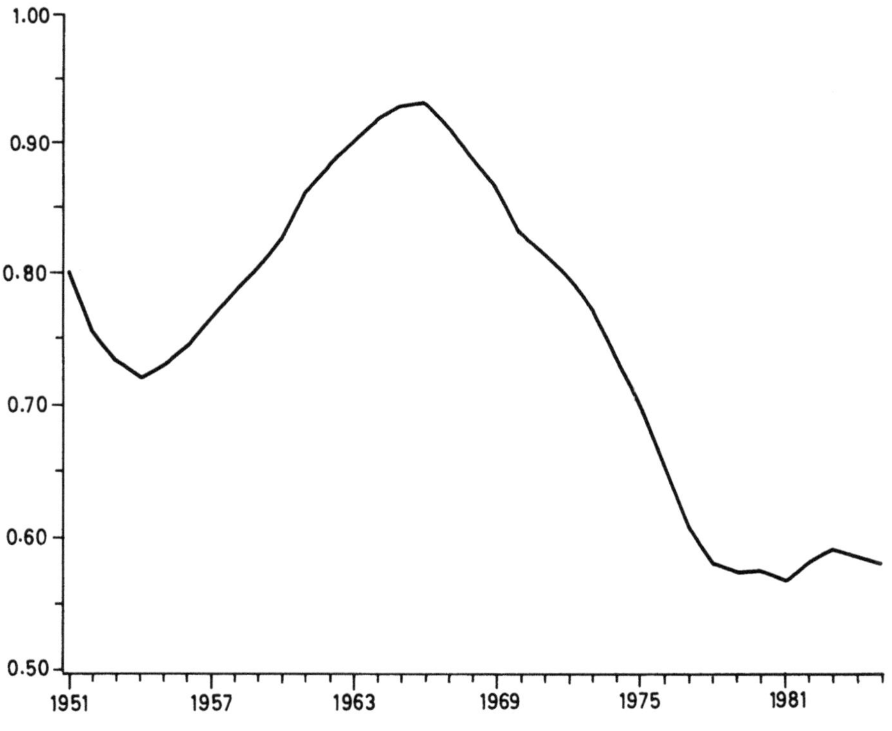

Fig. 4.5 Child dependency ratio

unemployment. Equation 4.11 includes a term (SC) which is hypothesized to influence structural unemployment along lines suggested by Layard and Nickell (1985). Such unemployment forms part of equilibrium unemployment. How SC is proxied is described in Section 4.4.

4.2.3 Wage behaviour

Our basic hypothesis is premised upon the augmented Phillips curve, i.e.

$$\Delta \ln W = \pi^e + \Delta \ln w^* + F_3(U - U^*) \tag{4.12}$$

where π^e is expected inflation, w^* is the equilibrium real wage and $U = (\bar{L} - L^D)/\bar{L}$. Equation 4.12 states that when the rate of unemployment equals the 'natural' rate expected real wage growth is equal to equilibrium real wage growth. The latter is defined below. If, however, $U > U^*$ there is an excess supply in the labour market so that real wages grow at less than their equilibrium rate. The opposite happens when $U < U^*$.

In the short run the labour market does not clear because U does not necessarily equal U^*. We assume that, in the short run, employment is on the demand curve. If $U > U^*$ this assumption is straight forward. If $U < U^*$ it implies that in the short run workers are prepared to be off their supply schedule. However, the corollary of this departure is extra wage pressure to justify the extra labour that is being supplied. This approach to disequilibrium modelling differs from the 'short side of the market' methodology that is illustrated in Chapter 3. In the long run Equation 4.12 implies that the labour market clears. If for simplicity U^* is constant, $\Delta \ln w^* = 0$, $L_g = 0$, $\pi^e - \pi$ we may write Equations 4.12, 4.7 and 4.5 as

$$\begin{aligned} \frac{d \ln w}{dt} &= \alpha_0 + \alpha_1 (\ln L^D - \ln (1 - U^*)\bar{L}) \\ \ln \bar{L} &= \beta_0 + \beta_1 \ln w \\ \ln L^D &= \xi_0 - \xi_1 \ln w \end{aligned} \tag{4.13}$$

Equations 4.13 imply the following general solution for the behaviour of real wages

$$w(t) = A \exp[-\alpha_1(\beta_1 + \xi_1)t] + \frac{\alpha_0/\alpha_1 + \xi_0 - \beta_0 - \ln (1 - u^*)}{\xi_1 + \beta_1} \tag{4.14}$$

where A is an arbitrary constant. The particular integral in Equation 4.14 is the equilibrium real wage (w^*). Since the root of the model is negative in the long run we find that

$$(1 - U^*)\bar{L} = L^D$$

in which case U tends to equal U^*. Equation 4.14 implies that w^* varies directly with U^* and autonomous labour demand (ξ_0) and it varies inversely with autonomous labour supply (β_0). It also implies that the Phillips curve can be regarded as an 'error correction mechanism' which together with the specification of labour supply and demand determines the level of wages as well as their rate of change. Therefore, while it is true that Equation 4.12 does not tie down the level of wages on its own, it is not true that the system as a whole does not determine the level of wages.

For estimation purposes Equation 4.11 is substituted into Equation 4.12 and we estimate Equation 4.15:

$$\Delta \ln W = \pi^e + \Delta \ln w^* + F_4(\overset{-}{\tilde{U}}, B/\overset{+}{W}, B/\overset{+}{P_c}, \overset{+}{S}C) \tag{4.15}$$

By setting $\Delta \ln w = \Delta \ln w^*$ we may use Equation 4.15 to infer U^* in terms of the arguments in Equation 4.11.

4.2.4 Overview of the model

When $U = U^*$ the labour market is in long-run equilibrium in the sense that real wages will be at their natural levels (assuming price expectations are fulfilled). This situation is illustrated in Fig. 4.6. The D schedule is based upon Equation 4.5 while the total demand for labour is represented by D^*. The S schedule is based upon Equation 4.7 and reflects the participation ratio. S^* represents the effective supply of labour, the gap between S^* and S reflects 'natural' unemployment and therefore Equation 4.10. The labour market is in equilibrium at e and the equilibrium real wage is equal to w^*. The working population is OL_2 of which OL_1 are working. Thus $L_1L_2 = u^*$ are 'naturally' unemployed and OL_3 work outside the general government sector. The natural rate of unemployment, U^*, is thus U^*/L_2.

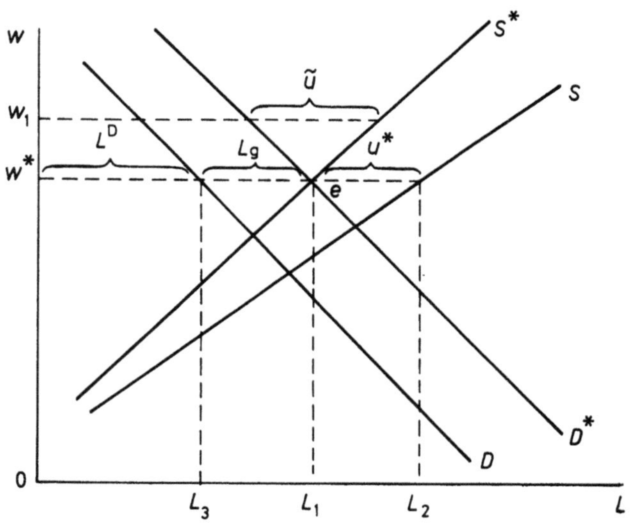

Fig. 4.6 The labour market

The schedules are drawn for given assumptions about tax rates, relative prices, etc. If, for instance, the population of working age (POP) rises, S will shift to the right and because U^* is unchanged S^* will shift in sympathy. This will lower w^* and raise L_1 and therefore L_3. If the 'natural' rate rises because, for example, benefits have been increased, S will remain unchanged but S^* will shift to the left; w^* will rise and L_1 and L_3 will fall. An example of a demand-side shock is a reduction in general government employment. If L_g falls D^* contracts, but D remains unchanged; w^* and therefore L_1 falls. However, in this case L_3 rises. Thus general government employment partially 'crowds out' private sector employment.

If real wages are above their equilibrium level at w_1 either because the schedules have shifted changing e, or because of a random shock to real wages, unemployment will rise. In addition to 'natural' unemployment there will be 'unnatural' unemployment represented by u in Fig. 4.6. According to Equation 4.12 this will induce a fall in real wages over time and the model will tend to e.

4.3 ECONOMETRIC RESULTS

In this section the equation specifications derived above are implemented using data drawn from the post-war period. Analogous estimates for the inter-war years are to be found in Beenstock and Warburton (1986), while Beenstock *et al.* (1986) documents an earlier attempt at explaining the post-war data in the context of the full CUBS model. Detailed data definitions and sources are contained in Appendix A. Variable symbols are consistent with those used in the previous section except for u, the unemployment rate, which becomes UR and for W, which there represented the cost of labour in general terms. Here, N stands for average gross earnings and W for average post-tax earnings. Furthermore, N^* represents the average basic gross hourly wage. Thus N and W reflect variations in overtime and short-time hours, but N^* does not. The linkages between N and W, together with all other interrelationships of variables are shown in Table 4.4. The four estimated equations relate to business sector employment (Equations 4.5 and 4.6) male labour supply Equation 4.8, female labour supply Equation 4.9 and the wage (earnings) adjustment Equation 4.15. Estimated single equation results for these are described in detail below.

4.3.1 Labour demand by the business sector

The specification of the labour demand equation shares many characteristics with those reported by Nickell and Andrews (1983) and, most recently, by Andrews (1987). Four differences remain between our work and theirs. First, Nickell and Andrews include in their dependent variable, government employment, whereas it is excluded in our work. Second, real materials and fuels prices are combined in the Nickell–Andrews equation but not in ours. Third, in this chapter, though not previously, the capital stock coefficient is constrained to unity, ensuring constant returns to scale. However, the restriction is tested and the results shown. Finally, the sample period used here extends to 1985 whereas Nickell and Andrews curtail theirs in 1979. As will be seen, the volatile years that followed 1979 provide a keen test of any UK labour market model.

In reaching a decision as to the most apposite empirical version of Equation 4.5, a choice of regressors was identified in two dimensions. First, whether to use the factor cost gross domestic product (GDP) deflator or the market price total final expenditure (TFE) deflator for P. In theory, P should be a gross output factor cost price. No gross output price index is available for the whole economy. The

Table 4.1 Business sector employment equations.
Dependent variable: $\Delta \ln (L/K)_t$. Sample period: 1953–1985

Equation:	*(4.16)* *GDP deflator*	*(4.17)* *TFE deflator*	*(4.18)* *GDP deflator*	*(4.19)* *TFE deflator*
DF	26	26	26	27
Constant	−3.598	−1.792	−2.419	−0.368
	(5.6)	(3.4)	(3.9)	(4.2)
$\ln (L/K)_{t-2}$	−0.484	−0.246	−0.321	−0.169
	(5.6)	(3.4)	(4.0)	(3.7)
$\ln (N(1+e)/P)_t$	−0.394	−0.210		
	(3.4)	(1.5)		
$\ln (N(1+e)/P)_{t-2}$	−0.315	−0.133		
	(2.8)	(1.2)		
$\ln (N^*(1+e)/P)_t$			−0.135	—
			(1.3)	
$\ln (N^*(1+e)/P)_{t-2}$			−0.253	−0.185
			(3.1)	(3.5)
$\ln (P_m/P)_t$	0.127	0.142	0.135	0.193
	(3.5)	(2.8)	(2.9)	(4.4)
$\Delta \ln (P_e/P)_t$	−0.061	−0.101	−0.057	−0.106
	(1.3)	(1.9)	(1.1)	(2.0)
$\ln (P_e/P)_{t-1}$	−0.028	−0.011	−0.036	−0.012
	(2.1)	(0.7)	(1.9)	(0.9)
$\bar{R}^2$	0.684	0.536	0.604	0.561
SE	1.01%	1.22%	1.13%	1.19%
Z_1	0.37	0.92	0.37	0.57
LM(1)	1.71	0.86	0.22	0.09
LM(2)	4.11	2.66	2.65	1.29
LM(3)	4.90	3.67	3.56	3.56
LM(4)	5.03	6.21	5.37	7.96
Chow (3, DF-3)	0.37	0.52	0.56	0.86
Chow (5, DF-5)	2.11	3.03	2.66	3.46
Chow (7, DF-7)	1.74	2.19	1.79	2.35

argument is over the relative merits of TFE, which approximates to a gross output definition, but is measured on a market price basis, and TPI, which is a factor cost deflator but refers to a net output concept. Second, whether to use average earnings (N) or hours adjusted earnings, (N^*). In both cases it seemed appropriate to employ a non-nested testing procedure since virtually all the regressors were affected by the measurement issue.

The estimated equations are summarized in Table 4.1 while the non-nested tests are contained in Table 4.2. No encouragement was offered in support of either the use of the TFE deflator or the hours-adjusted earnings variable. In Table 4.2, the coefficient on the fitted values from Equation 4.16 is insignificantly different from unity in each case. Moreover, this equation has more to commend it in terms of long-run properties and specification tests. The long-run real wage elasticity is just under 1.5, a result not dissimilar from that obtained by Symons (1985) for manufacturing. Following the discussion in Section 4.2, particularly that concerning Fig. 4.2, the coefficients on real energy and materials prices provide examples of long-run gross complementarity and gross substitutability, respectively, each with respect to labour.

To test the constant returns to scale restriction for the capital stock, a term in the logarithm of the stock was included in the regression. This t-test (Z_1) was failed easily. The other elasticities, for real materials and real energy prices, were both consistent with previous work. The autocorrelation properties were satisfactory to degree 4. Despite the high F-values for the predictive tests from 1978 and 1980, neither are significant at the 5% level. Thus specification 4.16 was

Table 4.2 Non-nested tests of employment equation specification

	(1) *Final expenditure* *Deflator*	*(2)* *Adjusted earnings* *Series*	*(3)* *Both*
Constant	−0.001	−0.003	0.0006
	(0.2)	(0.1)	(0.1)
Fitted values	1.124	1.089	0.912
from Equation 4.16	(3.8)	(2.8)	(3.7)
Fitted values	−0.145		
from Equation 4.17	(0.4)		
Fitted values		−0.096	
from Equation 4.18		(0.2)	
Fitted values			0.106
from Equation 4.19			(0.4)
LM(1)	1.94	1.63	1.54
LM(2)	3.35	3.06	2.95
LM(3)	3.66	3.53	3.43
LM(4)	3.77	4.00	4.07

retained pending consideration of the role of additional demand terms, as discussed in Section 4.2.

The decision whether or not to depart from the strict confines of the neoclassical model and to allow variables such as the real money supply and world trade volume to influence employment directly was made on the basis of a nested testing procedure. After a little experimentation, four demand terms appeared promising. These were the first difference of the yield on long-dated gilt-edged securities, current and lagged two years, the growth of real money, lagged one year, and the current growth rate of world trade volume. Adding all these terms yielded an *F*-test value of 1.33 compared to the appropriate 5% critical value of 2.74. Only in the case where the most significant variable was retained did the augmented model prove statistically superior. In the light of this lukewarm evidence for independent demand terms it was decided to proceed with the 'pure' neoclassical factor demand equation.

4.3.2 Male labour supply

Macroeconomic studies of labour supply cannot expect to do justice to the theoretical microeconomic model. However, in one respect there is a uniform result. The real post tax wage for males (WM) appears to have no effect whatsoever on male participation. The only two arguments which we were able to identify are both proxy variables for well-known economic phenomena. In Warburton (1987) it is argued that as real total retirement benefits increase relative to real post-tax wages, more men will be persuaded to retire early. Ill health also plays an important part in this decision, but on its own cannot explain the lower rates of participation seen since 1974 for males aged 50 to 64. The variable RRMP is constructed as the ratio of males aged 50 to 64 to those aged 15 to 49 multiplied by the ratio of real pension income to real earned income. As relative pension incomes rise (for example, because the retired have greater entitlements to earnings-related company pensions) or as the size of the 50 to 64 year old cohort increases in relative terms, male participation is liable to fall.

The second argument is the structural change proxy (SC). It is defined as the absolute value of the annual percentage change in industrial output relative to total output. As the pace of structural change increases, any mismatch of skills with jobs is likely to become more acute. The old are less able to adjust and more likely to be attracted by the terms of an early retirement offer, since severance pay is usually related to length of service. Hence, the faster the process of industrial change, the lower the male participation rate.

In Equation 4.20 the logit form of the specification is used. This enables elasticities to vary according to the participation rate. For example, a 1% rise in RRMP would induce a 0.12% fall in male participation from 1960s level of 0.877, but a 0.24% fall from 1985s level of 0.763. In particular, no matter how great the change in a right-hand side variable, the predicted participation ratio cannot stray outside its (0, 1) limits.

$$\Delta \ln (\text{PRM}_t/(1-\text{PRM}_t)) = \underset{(5.0)}{-0.303} \underset{(3.8)}{-0.134} \ln (\text{PRM}_{t-2}/\text{PRM}_{t-2})) \underset{(2.4)}{-0.288} \ln \text{RRMP}_t \underset{(1.3)}{+0.153} \ln \text{RRMP}_{t-1} \underset{(1.5)}{-0.447} \Delta \text{SC}_t \underset{(1.6)}{-0.616} \text{SC}_{t-2} \quad (4.20)$$

Sample period = 1955–1985 $\bar{R}^2 = 0.524$
$SE^* = 0.0048$
DF = 25
$Z_2(3, 25) = 0.71$ LM (1) = 0.68
Chow (3, DF-3) = 2.57 LM (2) = 0.81
Chow (5, DF-5) = 1.44 LM (3) = 1.28
Chow (7, DF-7) = 1.26 LM (4) = 2.22
*Calculated at the mean participation rate

The most disappointing aspect of the equation is the severe under-prediction of the participation rate in 1985, which is reflected in the parameter stability tests. However, none of the *F*-values are significant at the 5% level. The autocorrelation properties appear to be very good.

4.3.3 Female labour supply

The purpose of disaggregating labour supply by sex is to draw the distinction between the wage-inelastic behaviour of males and the wage-elastic behaviour of females. To a large extent this is an observation about the way the tax and benefit system discriminates between the sexes. In its simplest form, many women have little or no financial incentive to register their unemployment and hence disappear from the active labour force when they are not employed. The original specification of the female participation equation contains four arguments. Real post-tax female wages (WF/P_c), the structural change proxy (SC) which was introduced in the previous discussion, the lagged change in the unemployment rate (UR), and the child dependency ratio (CH). The unemployment rate is used to represent a discouraged worker effect, whilst the child dependency ratio stands for 'delayed motherhood' effect. Although this latter argument does not find empirical support here, a consideration of childbearing behaviour is clearly important to female labour supply. Over the past fifteen years it appears that couples have been delaying the birth of the first child longer than before, and have been reducing the length of time between the first and last birth. It is too soon to tell whether completed family size has also been falling. Counteracting these effects has been the increasing degree of provision of child-care facilities, both privately and by the state. This may help to explain the inconclusive results obtained.

The closest representation to the model specified in Equation 4.9 is detailed below:

$$\begin{aligned}\Delta \ln (\text{PRF}_t/(1-\text{PRF}_t)) = \underset{(5.0)}{0.070} - \underset{(3.0)}{0.233} \ln (\text{PRF}_{t-1}/(1-\text{PRF}_{t-1}))\\ - \underset{(2.7)}{0.015}\ \Delta \text{UR}_{t-1}\\ - \underset{(1.8)}{0.259} (\text{SC}_t + \text{SC}_{t-1} + \text{SC}_{t-2})\\ + \underset{(3.0)}{0.169} \ln (\text{WF}/P_c)_{t-2}\end{aligned} \tag{4.21}$$

Sample period: 1953–1985	$\bar{R}^2 = 0.331$
	$\text{SE}^* = 0.0098$
	$\text{DF} = 28$
	$\text{LM}(1) = 0.53$
Chow (3, DF-3) = 1.54	$\text{LM}(2) = 1.31$
Chow (5, DF-5) = 2.13	$\text{LM}(3) = 3.78$
Chow (7, DF-7) = 2.29	$\text{LM}(4) = 7.93$
	*Calculated at the mean participation rate

The real wage elasticity, again calculated at the mean participation rate, is 0.32. This result is broadly in line with previous unpublished estimates made for the post-war years and with the inter-war elasticity reported in Beenstock and Warburton (1986), (0.52). As the equation was satisfactory in respect of autocorrelation and parameter stability properties, and the other coefficients took plausible values, it was adopted for the model.

4.3.4 Earnings adjustment equations

Since the level of earnings is determined elsewhere in the model through the interaction of the demand and supply schedules, equation specification 4.15 is used to 'solve' for the unobservable variable, UR*. As real benefits (B/P_c) and structural change (SC) move UR* up and down, so a given observed rate of unemployment, UR, will have a variable impact on the growth of earnings. The size of the coefficient on actual unemployment determines whether the impact on earnings growth of a given change in (UR – UR*) is large or small.

The role of price expectations in the specification is important in two respects. First, it must be decided whether or not agents are to be credited with the knowledge of the *actual* average price level in year *t* when bargaining for earnings in year *t*. In the case where agents are assumed to have perfect foresight, up to one year ahead, a tight-fitting equation can be expected, but with poor outside sample performance. However, if the reverse is assumed, the fit of the equation is likely to be markedly inferior. The second dilemma concerns inflation homogeneity. One of the hazards of estimating an equation over a 35-year period in which inflation varied between 0% and 30% is the degree of inflation illusion deemed admissible. While it may be unreasonable to suggest that agents are fooled by inflation in 1987, it may have been a less ridiculous proposition in 1957. Certainly, in our inter-war estimation there was a strong indication of non-homogeneity, i.e.

$$\frac{\Delta \ln N}{\Delta \ln P} = \alpha \quad \text{where } \alpha < 1$$

Rather than explore the problematic route of variable parameter models, it was decided to test for inflation homogeneity in the usual way, but nonetheless to assert our prior belief that α should be unity. The results are contained in Table 4.3.

Table 4.3 Earnings equations.
Dependent variable: $\Delta \ln N_t$. Sample period: 1953–1985

Equation:	*4.22 Homogeneous, expectational*	*4.23 Unconstrained, expectational*	*4.24 Homogeneous, actual*	*4.25 Unconstrained, actual*
DF	25	23	25	24
Constant	−0.350	−0.390	−0.093	−0.230
	(2.9)	(3.2)	(1.2)	(3.5)
$\Delta \ln \text{PEXP}_t$	0.4*	0.413		
		(3.6)		
$\Delta \ln P_t$			1.0*	0.749
				(14.5)
$\Delta_3 \ln P_{t-1}$	0.2*	0.095		
		(1.8)		
$\Delta \ln \text{RER}_{t-2}$	−0.233	−0.197	−0.025	−0.065
	(2.8)	(2.5)	(0.4)	(1.6)
UR_t	−0.0082	−0.0064	−0.0028	−0.0036
	(4.3)	(3.0)	(2.2)	(3.7)
$\ln (B/P_c)_t$	0.087	0.096	0.027	0.060
	(3.0)	(3.4)	(1.4)	(3.8)
SC_t	0.566	0.801	0.147	0.490
	(1.9)	(2.7)	(0.7)	(3.0)
SC_{t-2}	0.735	0.850	0.236	0.430
	(2.2)	(2.8)	(1.1)	(2.6)
IP_t	−0.045	−0.028	−0.022	−0.017
	(2.7)	(1.7)	(1.9)	(2.0)
ΔIP_t	0.027	0.021	0.008	0.012
	(1.7)	(1.4)	(0.8)	(1.6)
SE	2.34%	2.15%	1.58%	1.15%
LM(1)	0.17	0.36	0.69	0.00
LM(2)	3.87	5.59	2.16	8.91
LM(3)	3.84	5.98	3.63	9.98
LM(4)	4.41	8.82	4.02	15.6
Chow (3, DF-3)	1.72	1.20	1.27	0.51
Chow (7, DF-7)	2.51	1.39	3.24	1.38

* Constrained coefficients.

Equation 4.22, which was estimated in canonical form, restricts the coefficients on inflation to be unity, yet employs one-step ahead price expectations derived from the CUBS model. The result is a poor-fitting equation with good autocorrelation but only mediocre parameter stability properties. Nevertheless the individual parameters are well determined and correctly signed. To test inflation homegeneity, Equation 4.23 was estimated. The calculated *F*-statistic is 2.76 against a critical

Table 4.4 Labour market identities

Equations		
L_f	$L + \overline{LX} - \overline{L_m}$	
LSUPF	$PRF \cdot POPF1$	
LSUPM	$PRM \cdot POPM1$	
NF	$0.005688 \cdot FMWD \cdot NM$	
NM	$(TWB - EF \cdot L_f)/\overline{L_m}$	
TWB	$N(L + \overline{LX})$	
U	$LSUPM + LSUPF - L - \overline{LX}$	
UR	$(U/WPOP)\ 100$	
WF	$(NF - \overline{TF}/1000)(1-\overline{MTR}-\overline{MNIR}) + \overline{TF}/1000$	
WM	$(NM - \overline{TM}/1000)(1-\overline{MTR}-\overline{MNIR}) + \overline{TM}/1000$	
WPOP	$L + \overline{LX} + \overline{LSE} + U$	(4.26)
Definitions		
FMWD	Female to male wage differential	
L	Business sector employees in employment	
L_f	Female employees in employment	
L_m	Male employees in employment	
LSE	Self-employment	
LSUPF	Female labour supply	
LSUPM	Male labour supply	
LX	Other employees in employment	
MNIR	Contracted-in rate of national insurance contributions	
MTR	Standard rate of income tax	
N	Average annual gross earnings	
NF	Average annual female gross earnings	
NM	Average annual male gross earnings	
POP_f	Female population of working age	
POP_m	Male population of working age	
PRF	Female participation ratio	
PRM	Male participation ratio	
TF	Married woman's or single person's annual tax allowance	
TM	Married man's annual tax allowance	
TWB	Total wage bill for employees	
U	Unemployment (numbers)	
UR	Unemployment rate (%)	
WF	Average annual post-tax female earnings	
WM	Average annual post-tax male earnings	
WPOP	Working population	

Note: A bar above a variable denotes that it is treated as exogenous to the model.

5% value of 3.38. Although it is possible, therefore, to maintain the hypothesis that $\alpha = 1$, the calculated statistic is uncomfortably close to significance. While the non-homogeneous equation has a greater degree of parameter stability, autocorrelation is more of a problem.

Comparing Equation 4.22 with Equation 4.24, in which the actual price change is used rather than the expectation, a massive improvement in the standard error is revealed. When homogeneity is relaxed, as in Equation 4.25, a further gain in efficiency occurs. The *F*-test rejects homogeneity this time and by a very large margin. However, the cost is a severe deterioration in autocorrelation beyond the first order.

The multi-dimensional nature of the estimation problem makes it difficult to arrive at a preferred equation. In the exercises that follow, the worst fitting equation, 4.22, has been employed. Its advantages lie in the fact that it has the strongest unemployment term of all four equations and also possesses reasonable implied responses for real benefits and the structural change proxy.

4.3.5 Other model equations

Table 4.4 lists the identities and technical relationships required to complete the model. Summary variable definitions are provided but full descriptions are relegated to Appendix A.

4.4 SIMULATIONS

The model described by Equations 4.16, 4.20, 4.21 and 4.22 and the block 4.26 was subjected to a series of simulation exercises. In the first part of this section the one step ahead errors are examined. A series of six dynamic simulation exercises follows. Finally, we attempt to decompose the rise in unemployment between 1972 and 1985.

4.4.1 One-step-ahead simulation

Absolute and absolute percentage errors for the one-step-ahead (OSA) model predictions over the period 1956–1985 are given in Table 4.5. Figures 4.7 to 4.9 give more detail for employment, the unemployment rate and earnings. The earnings results are shown in the form of percentage changes only because the differences in earnings levels are barely distinguishable on a graph. The model determines the level of earnings as well as the change. The only blemishes in the OSA tracking record are in 1967 for business employment, and between 1976 and 1979 for earnings. This latter volatility is believed to be induced by inflation uncertainty during the period in which monetary targets were adopted in the UK. Figure 4.8 suggests that the errors derived for the unemployment rate may be autocorrelated. This proposition was put to the test and the first four terms of the unemployment

Table 4.5 One-step-ahead errors

	Annual absolute error	*Annual absolute % error*
Business sector employment ('000)	125	0.70
Male labour supply ('000)	53	0.37
Female labour supply ('000)	67	0.75
Unemployment ('000)	125	0.48*
Earnings (£p.a.)	32	1.10

*Error in the percentage unemployment rate.

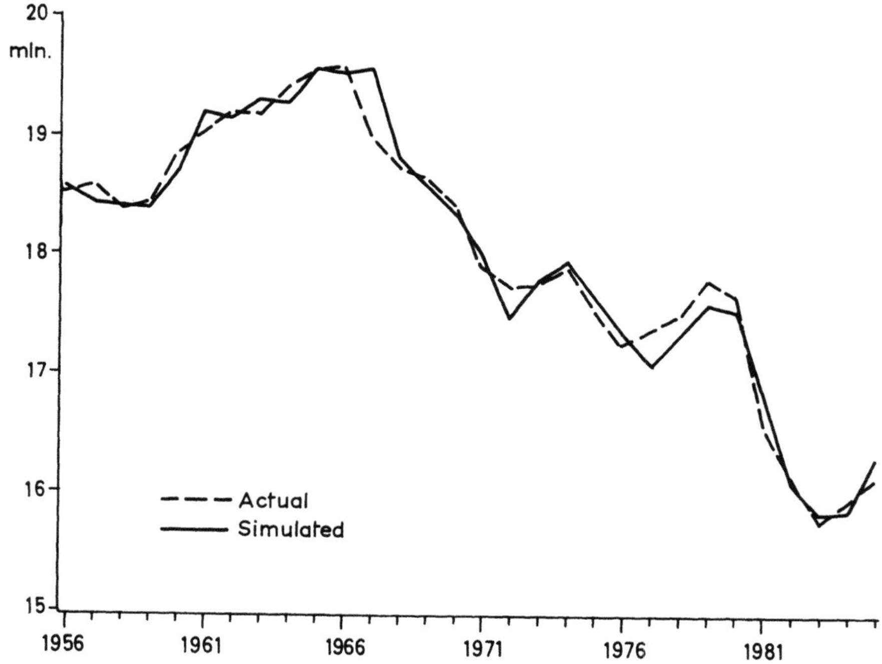

Fig. 4.7 Employment: static

error correlogram were found to be 0.312, −0.015, −0.127 and 0.061. The Ljung-box test statistic was 5.81 and therefore within the initial 5% significance bound. Accordingly the hypothesis of non-randomness is rejected.

Table 4.6 reports the vector autocorrelation matrix to degree 4 for the OSA simulation exercise. The 95% significance level is reached by a correlation (of absolute size) greater than 0.39. Only four such correlations arise (adjacent diagonal correlations count only once) out of 70 unique pairings. The failure rate of 5.7% agrees well with the chosen level of significance, leading us to reject the notion

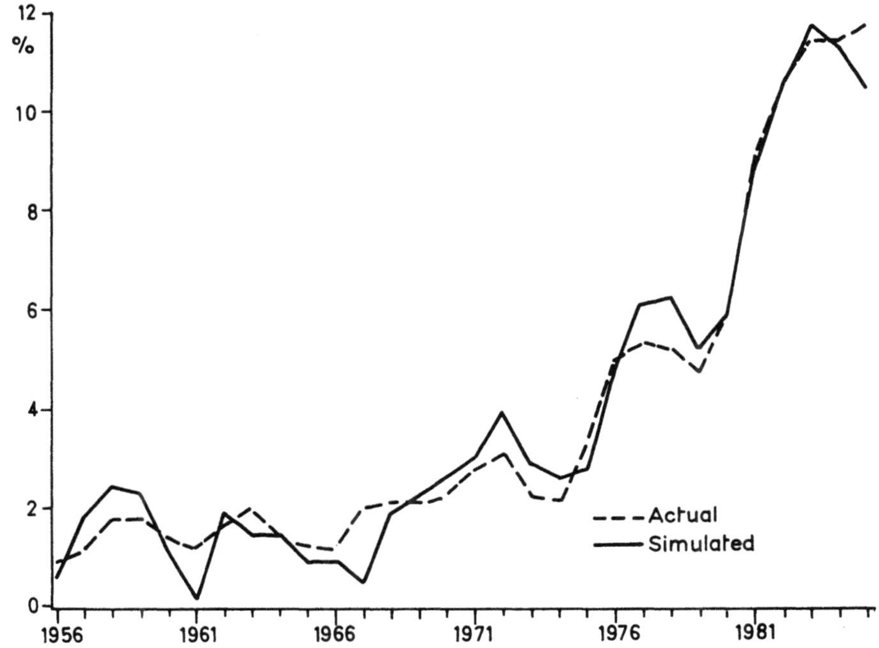

Fig. 4.8 Unemployment rate: static

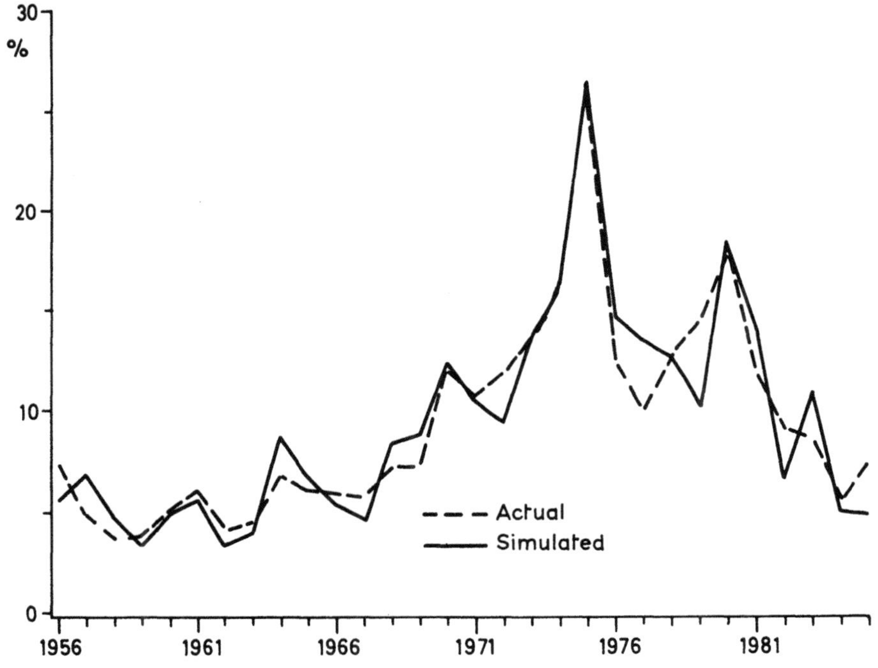

Fig. 4.9 Earnings growth: static

Table 4.6 Vector autocorrelation matrix for one-step-ahead errors

	L^*_t	L^*_{t-1}	L^*_{t-2}	L^*_{t-3}	L^*_{t-4}	$LSUPM^*_t$	$LSUPM^*_{t-1}$	$LSUPM^*_{t-2}$	$LSUPM^*_{t-3}$	$LSUPM^*_{t-4}$	$LSUPF^*_t$	$LSUPF^*_{t-1}$	$LSUPF^*_{t-2}$	$LSUPF^*_{t-3}$	$LSUPF^*_{t-4}$	N^*_t	N^*_{t-1}	N^*_{t-2}	N^*_{t-3}	N^*_{t-4}
L^*_t	1																			
L^*_{t-1}	−0.05	1																		
L^*_{t-2}	0.01	−0.04	1																	
L^*_{t-3}	−0.26	0.01	−0.04	1																
L^*_{t-4}	−0.02	−0.26	0	−0.04	1															
$LSUPM^*_t$	0.21	0.04	−0.01	0.25	−0.17	1														
$LSUPM^*_{t-1}$	−0.29	0.33	0	0.03	0.26	0.03	1													
$LSUPM^*_{t-2}$	0.07	−0.29	0.33	−0.02	0.03	−0.11	0	1												
$LSUPM^*_{t-3}$	−0.24	0.02	−0.28	0.33	0	0.07	0.06	0.02	1											
$LSUPM^*_{t-4}$	−0.12	−0.22	0.01	−0.25	0.32	−0.01	0.02	0.04	0.05	1										
$LSUPF^*$	0.57	−0.28	0.06	−0.28	0.01	0.34	−0.28	0.20	−0.04	0.42	1									
$LSUPF^*_{t-1}$	−0.26	0.59	−0.29	0.06	−0.29	0.28	0.31	−0.27	0.26	−0.06	−0.23	1								
$LSUPF^*_{t-2}$	−0.15	−0.23	0.57	−0.31	0.05	−0.20	0.13	0.33	−0.18	0.18	−0.10	−0.23	1							
$LSUPF^*_{t-3}$	−0.09	−0.07	−0.22	0.55	−0.31	0.20	−0.18	0.14	0.31	−0.18	−0.11	−0.08	−0.16	1						
$LSUPF^*_{t-4}$	0.00	0.05	−0.20	−0.20	0.54	−0.33	0.11	−0.20	0.23	0.31	−0.10	−0.16	−0.18	−0.15	1					
N^*_t	−0.19	−0.12	0.24	0.26	−0.27	0.37	−0.24	−0.09	0.14	0.12	0.05	0	0.38	−0.25	1					
N^*_{t-1}	−0.37	−0.12	−0.19	0.32	0.26	−0.11	0.15	−0.31	0.08	0.12	−0.46	0.19	−0.15	0.05	0.33	−0.03	1			
N^*_{t-2}	−0.12	−0.40	−0.12	−0.20	0.34	−0.15	−0.03	0.16	−0.39	0.10	−0.09	−0.44	0.24	−0.17	0.09	−0.34	0.07	1		
N^*-3	0.06	−0.18	−0.40	−0.15	−0.19	0.09	0.02	−0.01	0.06	−0.38	0.12	−0.05	−0.34	0.22	−0.11	0.18	−0.21	0.01	1	
N^*_{t-4}	0.24	0.12	−0.22	−0.38	−0.18	−0.23	−0.08	−0.01	0.10	0.05	0.12	0.07	−0.19	−0.33	0.17	0.33	0.04	−0.17	0.12	1

Standard error $= 1/\sqrt{(26-1)} = 0.2$.

that cross-equation errors are non-random. This is a satisfying result given that only single equation estimation methods were used.

4.4.2 Dynamic simulations

(a) A 5p cut in personal income tax

Figure 4.10 shows the dynamic multipliers for a 5p tax cut with respect to employment ('000s), unemployment ('000s) and average earnings (%). It will be remembered that only female labour supply responds to the post-tax real wage and hence all the action originates here. The tax cut shifts the supply curve to the right raising employment by about 100 000 and reducing earnings by about 1/2%, eventually. Unemployment is almost unchanged as a result.

(b) A 1% increase in the net capital stock of the business sector

As can be seen from Fig. 4.11, a higher capital stock induces a rightward shift in labour demand, stimulating both employment and real earnings, initially. The oscillatory adjustment process (arising from the second order lag process in (L/K))

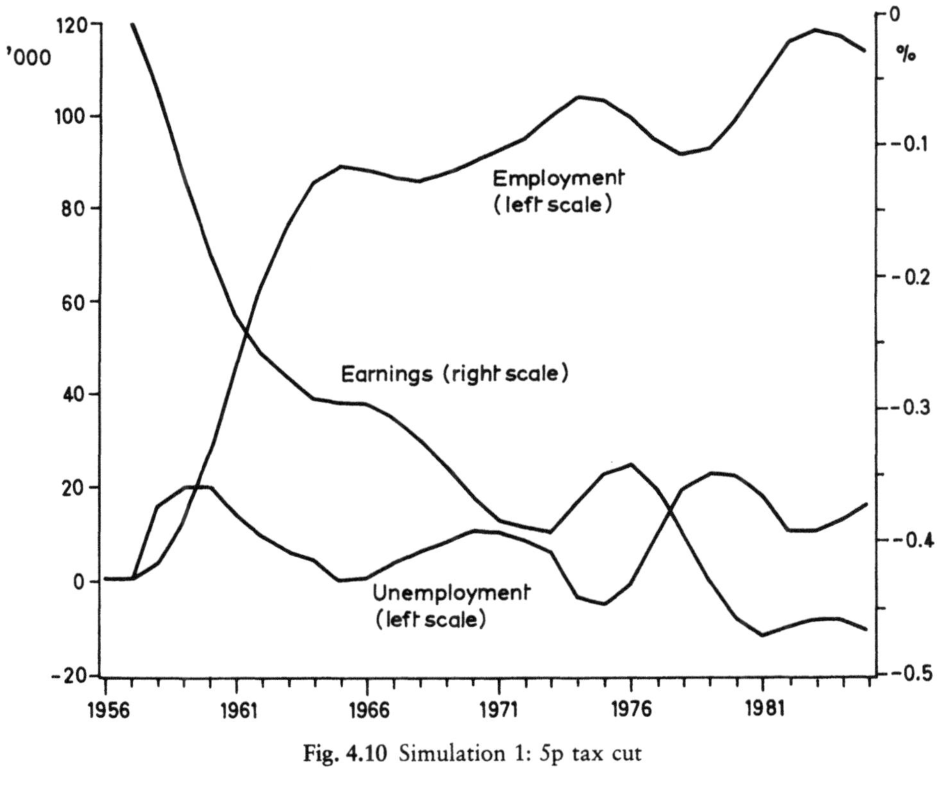

Fig. 4.10 Simulation 1: 5p tax cut

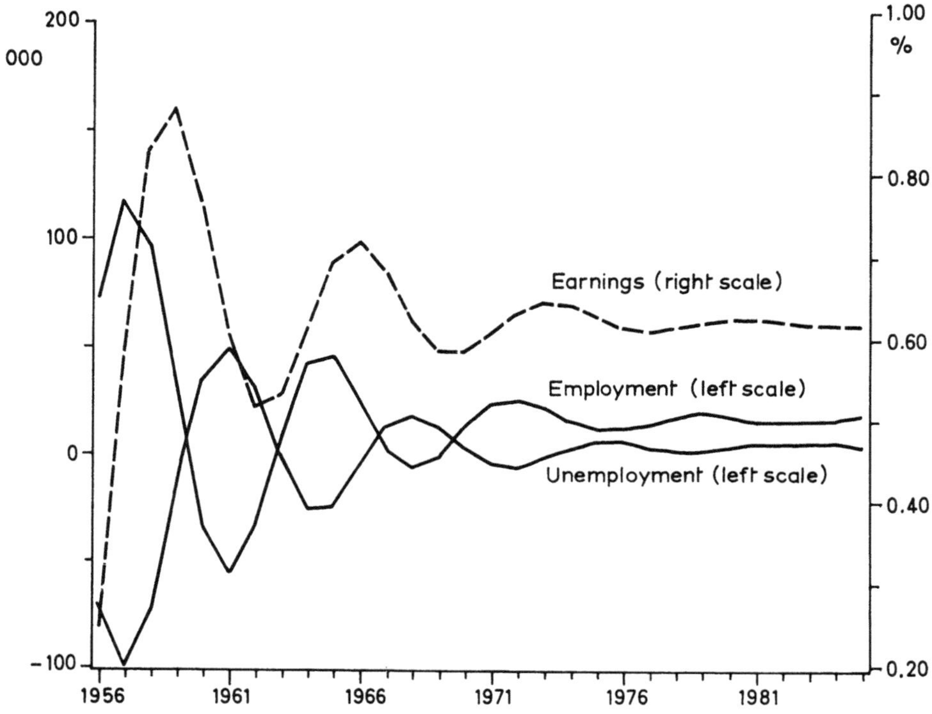

Fig. 4.11 Simulation 2: 1% increase in capital stock

dampens relatively quickly, leaving employment slightly higher and earnings over 0.6% stronger. If this is an accurate reflection of the UK business sector labour market, it implies that most of any gain in total factor productivity is distributed to the existing workforce in the form of higher pay.

(c) Faster pace of structural change

The structural change variable, SC, assumed an average value of 1.52% per annum between 1953 and 1972. Although volatile from year to year, it is always positive by design. The shock was to increase SC by 1% throughout. Two effects were identified in the estimation of the model (Fig. 4.12). First, structural change gives rise to structural unemployment. People who are structurally unemployed cannot, in most instances, expect to be re-employed in the same industry at the same skill level. During the period of retraining, or waiting to grow old enough to retire, the structurally unemployed do not exert downward pressure on the real wage.

The second effect, present in both male and female supply functions, is a recognition of the selective impact of structural change on the older members of the workforce, leading to non-participation. This represents a leftward shift in

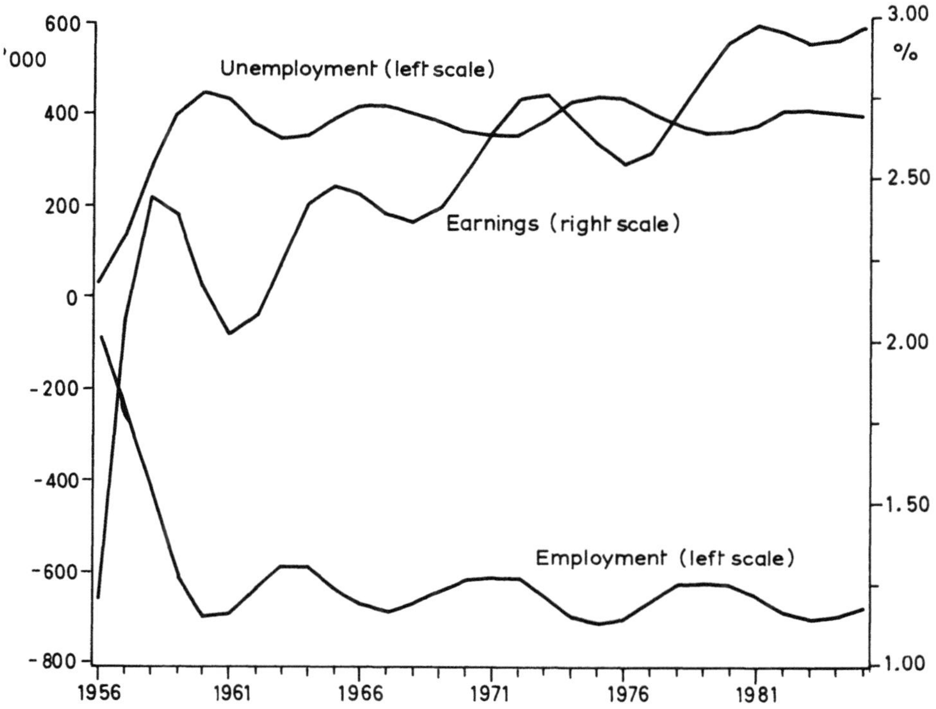

Fig. 4.12 Simulation 3: more structural change

the labour supply which reduces employment and raises earnings. Hence unemployment rises by about 400 000, but employment falls by more, about 700 000, as older people withdraw from the labour market. Earnings rise sharply since there has been a reduction in labour market competition (average number of unemployed applying for a given vacancy) and a contraction of labour supply.

(d) A 5% increase in the real value of benefits

This simulation, pictured in Fig. 4.13, is relatively straightforward to interpret. An increase in 'voluntary' unemployment of about 140 000 occurs as employment falls by an equivalent amount. Earnings rise by 0.5% in order to bring this about.

(e) A 10% drop in government employment

This simulation (Fig. 4.14) was described in theoretical terms using Fig. 4.6 in Section 4.2. By releasing 400 000 to 500 000 staff a depressing influence is exerted on the market, leading to a 2% cut in average earnings. This stimulates business employment and, after about 3 years, the 260 000 who were initially jobless have

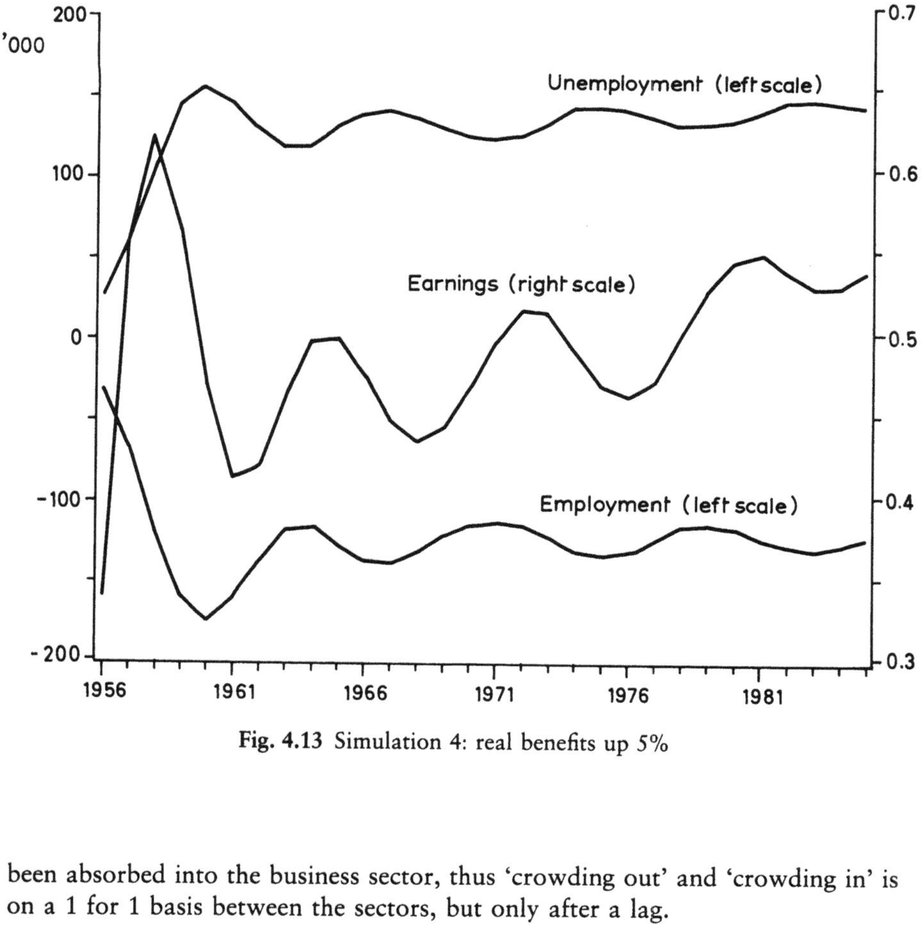

Fig. 4.13 Simulation 4: real benefits up 5%

been absorbed into the business sector, thus 'crowding out' and 'crowding in' is on a 1 for 1 basis between the sectors, but only after a lag.

(f) A 10% increase in relative female pay

As in Fig. 4.11, the simulation portrayed in Fig. 4.15 relies wholly on the female labour supply equation for its effect. The resulting paths for employment and unemployment came as something of a surprise. Rather than 'pricing themselves out of the market', females raise their participation level and force the whole economy wage level to fall by 0.6%. This allows employment to rise by almost 150 000 while unemployment rises only very slightly. Some tentative support is thereby provided for Tzannatos' (1987) descriptive thesis concerning the quality effect in female employment. As women compete for jobs in a wider variety of occupations on a full-time basis, a rerating of the female-to-male wage differential takes place. Superficially, female labour looks cheap relative to male labour and therefore more females are employed, up to the point where the rerating is complete.

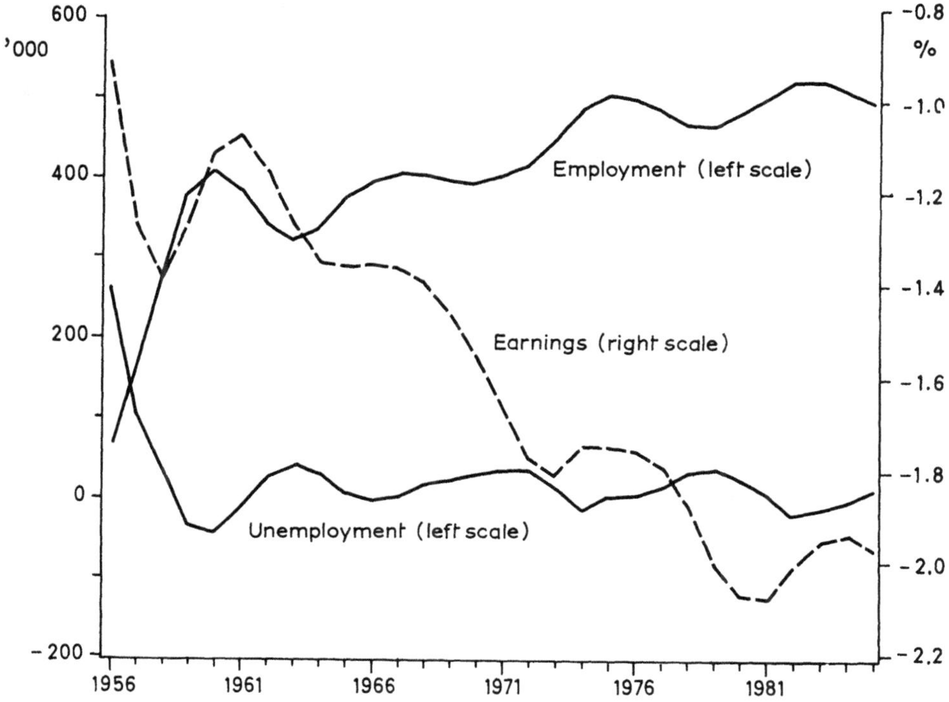

Fig. 4.14 Simulation 5: government employment down 10%

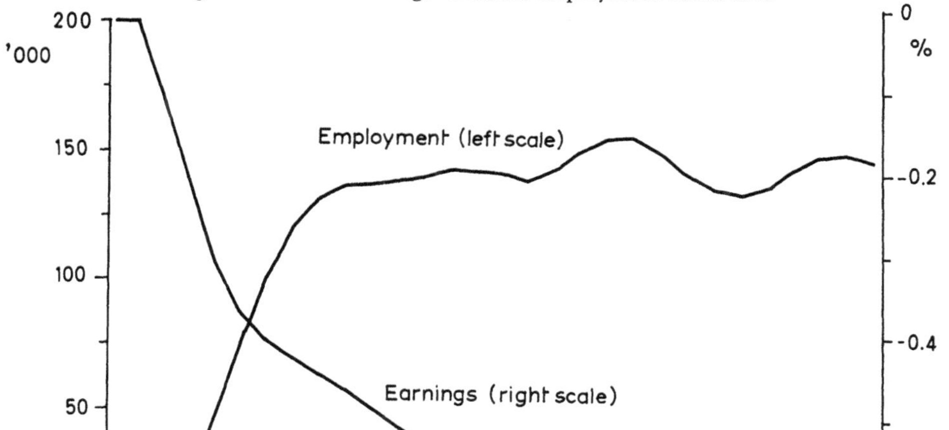

Fig. 4.15 Simulation 6: female pay up 10%

4.4.3 Explaining the rise in UK unemployment 1972–1985

The final dynamic simulation exercise concerns the topical issue of the causes of the current high unemployment. Figure 4.16 reveals that the model does not track unemployment at all well in the period 1962 to 1973. Furthermore, it anticipates the surge in unemployment in 1977–78 before the event in 1980–81. Much of this volatility can be traced to the (inaccurate) inflation expectations with which consumers were credited for the purposes of Equation 4.22. Even when the model is asked to forecast only from 1973, a similar pattern emerges. Therefore the stylized exercise which follows must be viewed in the light of these dynamic simulation properties.

In Table 4.7 a series of counterfactual statements are listed beneath the actual unemployment averages for the periods, 1973–76, 1977–80 and 1981–85. If the coefficients of the model are well determined and if the structure of the model is valid, then this table should yield potentially valuable information on the causes of unemployment. It is clear that no single item can account for the bulk of the explanation. Several are required to come even close to an adequate decomposition.

The contributions of the various elements differ between periods, but, structural change, real benefits and the net capital stock appear to be the most influential

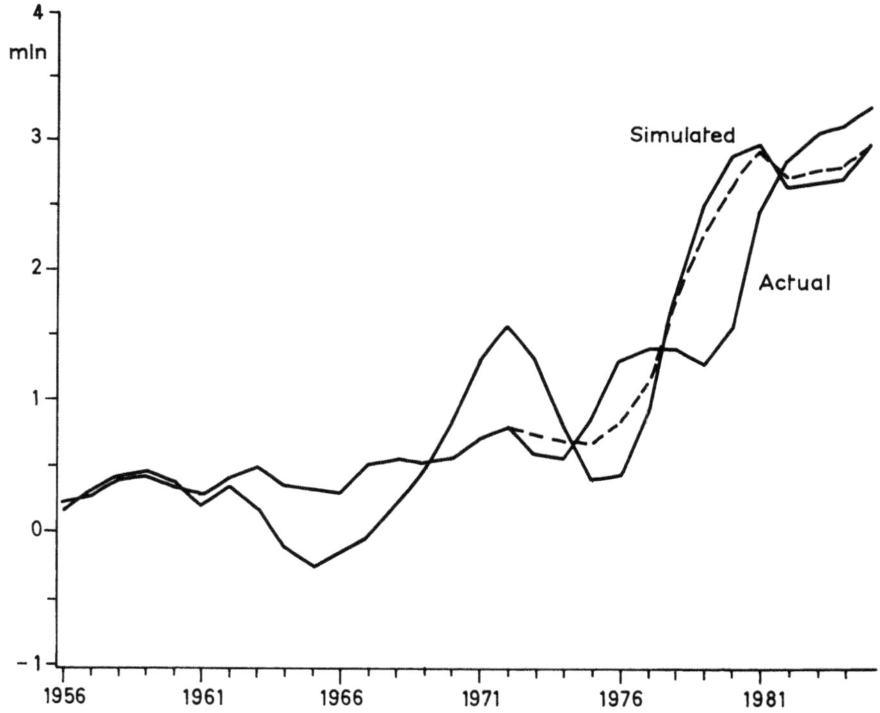

Fig. 4.16 Unemployment: actual and simulated

Table 4.7 Stylized decomposition of unemployment change, 1973–1985

Row	*Thousands, adult, seasonally adjusted excluding school leavers*	*1973–1976*	*1977–1980*	*1981–1985*
1.	Actual unemployment	797	1363	2872
2	Real energy prices falling by 2.2% per year	−155	−90	−31
3.	Real materials prices remaining at 1972 level	+433	−218	−8
4.	No increase in the population of working age after 1972	−102	−220	−266
5.	Structural change proceeding after 1972 at 1953–1972 average rate	−50	−62	−711
6.	Capital stock increasing after 1972 at 1953–1972 average rate	−190	−299	−503
7.	Real out-of-work benefits at 1966 level	−234	−326	−487
8.	Employers' labour taxes remaining at 1972 level	−46	−350	+91
9.	Employees' tax and N.I. rates remaining at 1972 level	−15	+3	−12
10.	Dynamic simulation errors	−90	+546	−115
11.	Other effects and unexplained residual	+102	+103	−380
12.	Average unemployment (1953–1972)	450	450	450

variables. Whilst the real value of benefits was raised as a deliberate act of policy in 1967, the other variables were beyond the control of any government. It is tempting to conclude, regarding row 2 of Table 4.7, that real energy prices played only a minor role in the unemployment story. However, this would be to ignore the indirect effects of real energy prices on the capital stock and on the size of the manufacturing industry. Reduced rates of profitability brought a sharp reduction in net capital formation, and the contraction of the manufacturing sector is picked up in our structural change proxy variable. Thus the total effect of the change in real energy and, especially, oil prices in 1973–74 and again in 1979–80 was almost certainly the prime cause of higher unemployment in this analysis.

Demographic factors and taxes appear to have played a supporting role in the explanation, but the alternative values proposed for these variables are deliberately advantageous to our point of view. If some modest increase in population of working age were allowed then the resulting impact on unemployment would be less. However, warts and all, the model provides an illuminating insight into the workings of the UK labour market and one which does not rely on a blanket 'shortage of aggregate demand' tag. Real energy prices did have severe demand effects but these cannot form the basis of a long-term answer to the unemployment question. The supply-side effects of the oil shock were longer lasting and, to some extent, cumulative. This is the nature of the problem facing the UK government in 1987: whether to undertake a capital-intensive spending program to compensate for the post-1972 business sector shortfall, or whether to wait for industry to feel confident enough about the future to carry through investment plans of their own.

APPENDIX A ADDITIONAL DATA

B	Real out-of-work benefits. Supplementary benefit scale rate for a childless couple plus half the sum of rent and rates, deflated by the Retail Price Index. Source: Whitbread, C., November 1986 New data series for benefits, mimeo, City University Business School.
CH	Ratio of numbers of dependent children aged 0 to 5 years to females aged 20 to 34. Source: *Annual Abstract of Statistics.*
$(1+e)$	Ratio of employers' superannuation and national insurance contributions to total wages and salaries, indexed to 1980 = 1. Source: *Economic Trends Annual Supplement.*
FMWD $(=\theta)$	Ratio of female to male hourly earnings for full-time employees in all occupations, indexed to 1970 = 100. Data refer to April each year. Source: *Employment Gazette.*
IP	An incomes policy dummy graded according to the severity of the policy. Assumes the values 1961 = 0.5, 1962 = 1.0, 1963 = 0.25, 1965 = 0.5, 1966 = 1.0, 1967 = 0.5, 1972 and

1973 = 0.25, 1975 = 0.5, 1976 = 1.0, 1977–1979 = 0.5, elsewhere zero. Source: Whitbread, C., November 1986, Modelling the union mark-up, mimeo, City University Business School.

K — Net capital stock of the business sector at constant 1980 replacement cost in £b. Calculated as the addition of the net stocks of agriculture, forestry and fishing, energy and water, manufacturing, construction, distribution, transport and communications, banking, insurance and finance, plus 0.14 of the net stock of other services. Source: Direct from the Central Statistical Office.

L — Employees in employment, thousands. Source: *Economic Trends Annual Supplement.*

L_f — Female employees in employment, thousands. Sources: *Employment Gazette, Historical Supplement* 1 and *British Labour Statistics Historical Abstract.*

L_m — Male employees in employment, thousands. Source: As above.

LSE — Self-employment, thousands. Source: As above.

LSUPF — Female employees in employment plus female unemployment, thousands. Source: Calculated from components.

LSUPM — Male employees in employment plus male unemployment, thousands. Source: Calculated from components.

LX — Employment in the non-business sector (General government and North Sea oil), thousands. Sources: (1961 to date) *Economic Trends Annual Supplement*, (1948–1960) estimated from *British Labour Statistics Historical Abstract.*

MNIR — Marginal rate of national insurance contribution for Class 1 contracted-in employees, between lower and upper earnings limits. Before 1975 this is approximated by the average adult National Insurance contribution. Source: *Annual Abstract of Statistics.*

MTR — Standard rate of personal income tax, pence per £. Source: *Inland Revenue Statistics.*

N — Average annual earnings, £000s. Total wages and salaries of employees in £b divided by total employees in employment in thousands. Source: Calculated from components.

N* — Average earnings (*N*) adjusted for hours worked. The adjustment is calculated as:

$$\frac{1.3\ (45) - 0.3\ (\text{Normal hours per week})}{1.3\ (\text{Actual hours per week}) - 0.3\ (\text{Normal hours per week})}$$

where 1.3 is the assumed overtime premium rate. Source: Nickell and Andrews (1983).

NF — Average annual female earnings, £000s. Calculated as (.005688). FMWD. NM.

NM — Average annual male earnings, £000s. Calculated as [TWB/(LM + LF (.005688) FMWD)].

P (GDP defl.) — Deflator for Gross Domestic Product at factor cost, 1980 = 100. Source: *Economic Trends Annual Supplement.*

P (TFE defl.) — Deflator for Total Final Expenditure at market prices, 1980 = 100. Source: *Economic Trends Annual Supplement.*

P_c — Retail price index, all items rescaled to 1980 = 100. Source: *Economic Trends Annual Supplement.*

P_e — Price of energy to the industrial sector. Constructed as a weighted average of solid fuel, gas, petroleum and electricity prices and scaled to 1980 = 100. Sources: *Department of Energy Digest of UK Energy Statistics and Energy Trends.*

P_m — Price of non-energy raw materials. Constructed as (0.6453 Pnm + 0.3547 Pimo) where:

$$\text{Pnm} = \frac{\text{Pimi} - 0.037\text{Pie}}{1 - 0.037}$$

Pimi = wholesale input price index for manufacturing industry, 1980 = 100.

Pimo = wholesale output price index for manufacturing industry, 1980 = 100.

Pie = unit value index of fuel imports, 1980 = 100.

Sources: *Economic Trends Annual Supplement and Monthly Digest of Statistics.*

PEXP — One-step ahead predictions for P (GDP deflator) from CUBS model simulation, 1980 = 100.

POP_f — Females aged 15 to 59 years, thousands. Source: *Annual Abstract of Statistics.*

POP_m — Males aged 15 to 64 years, thousands. Source: *Annual Abstract of Statistics.*

PRF — Female participation rate. Constructed as LSUPF/POP_f1. Source: Calculated from components.

PRM — Male participation rate. Constructed as LSUPM/POP_m1. Source: Calculated from components.

RER — Real multilateral exchange rate index, 1980 = 100. Calculated using producer prices for 18 countries and weights from IMF's multilateral exchange rate model. Source: London Business School data bank.

RRMP — Relative retirement payments for males. Constructed as the ratio of males aged 50 to 64 to those aged 15 to 49, multiplied by the ratio of real pension incomes to real earned incomes. Source: *Annual Abstract of Statistics and Blue Book.*

SC — Structural change proxy. Constructed as the absolute value of

	the annual percentage change in the ratio of manufacturing output to the output measure of Gross Domestic Product. Source: *Economic Trends Annual Supplement.*
TF	Single person's or married woman's annual personal tax allowance, standardised to calendar year averages, in £. Source: Inland Revenue Statistics.
TM	Married man's annual personal tax allowance, standardised to calendar year averages, in £. Source: Inland Revenue Statistics.
TWB	Total wages and salaries of all employees in employment, £b. Source: *Economic Trends Annual Supplement.*
U	U.K. adult unemployment (excluding school leavers) in June of the year, measured in thousands. Sources: *Employment Gazette, Historical Supplement* 1 and *British Labour Statistics Historical Abstract.*
UR	Unemployment percentage rate, defined as (100U/(LSUPM + LSUPF)). Source: Calculated from components.
WF	Annual average post-tax female earnings. Computed as (NF − TF/1000) (1 − MTR − MNIR) + TF/1000. Source: Calculated from components.
WM	Annual average post-tax male earnings, computed as (NM − TM/1000) (1 − MTR − MNIR) + TM/1000. Source: Calculated from components.
WPOP	Working population, derived as (LSUPF + LSUPM). Source: Calculated from components.

REFERENCES

Andrews, M. J. (1987) The aggregate labour market: an empirical investigation into market-clearing for the UK, *Economic Journal*, Vol. 97, pp. 157–76, March.

Beenstock, M., Dalziel, A., Lewington, P. and Warburton, P. J. (1986) A macro-economic model of aggregate supply and demand for the U.K., *Economic Modelling*, Vol. 3, No. 4, pp. 242–68, October.

Beenstock, M. and Warburton, P. J. (1986) The market for labour in interwar Britain, Centre for Economic Policy Research Discussion Paper 105, August.

Bruno, M. and Sachs, J. (1985) *The economics of worldwide stagflation*, Basil Blackwell, Oxford.

de Cooman E., Ermisch, J. and Joshi, H. (1985) The next birth and the labour market: a dynamic model of births in England and Wales, Centre for Economic Policy Research Discussion Paper 37, January, London.

Layard, P. R. G. and Nickell, S. J. (1985) The causes of British unemployment, *National Institute Economic Review* 111, pp. 62–85, February.

Nickell, S. J. and Andrews, M. J. (1983) Unions, real wages and employment in Britain 1951–1979, *Oxford Economic Papers*, Vol. 35 (Supplement), pp. 183–206.

Symons, J. S. V. (1985) Relative prices and the demand for labour in British manufacturing, *Economica*, Vol. 52, pp. 37–50, February.

Tzannatos, Z. (1987) Female pay; has the State unshackled the market?, *Economic Affairs*, Vol. 7, 2, pp. 26–8, January.

Warburton, P. J. Labour supply incentives for the retired, (1987) Chapter 6 of *Work, welfare and taxation*, ed. by M. Beenstock, Allen and Unwin, London.

Zabalza, A. and Tzannatos, Z. (1985) The effects of Britain's anti-discriminatory legislation of relative pay and employment, *Economic Journal*, Vol. 95, pp. 679–99, September.

5

A new classical model of the labour market

PATRICK MINFORD

Most new classical models – for example Sargent (1976) for the US – contain no explicit labour supply and demand equations, but rather an aggregate output supply equation derived from some implicit equations for labour market and productive technology. This was true also of the Liverpool Model, a new classical model of the UK, until in 1980 it became obvious to me that there had to be a full and explicit analysis of unemployment whose behaviour was diverging markedly from its traditional relationship with output (Okun's law).

The basic idea of the resulting Liverpool labour market equations is that the market for labour clears through the uncontrolled non-unionized market, but there is a large union sector in which powerful unions set the wages of their members; those displaced by high union wages from this sector move into the non-unionized sector but the resulting fall in non-union wages causes some non-union workers to withdraw their labour when they compare their lower net wages with available unemployment benefits. On the demand side of the labour market firms are assumed to be competitive and to have a fairly standard production function; they demand the labour for which marginal product equals the real wage and produce whatever this labour can turn out given the optimal capital stock at ruling international costs of capital. International conditions alter the terms of trade and shift firms' demand curve for labour.

These ideas are fully described in the sections which follow. Section 5.1 outlines a stylized version of the open economy model to be used. Sections 5.2 and 5.3 deal with the labour market in detail, this being the major focus of the chapter. In Section 5.4 we present our estimates for the UK labour market equations and using them as part of the macroeconomic model draw out the implications for the UK natural rate. Section 5.5 answers a number of critics and leads to a set of concluding remarks.

5.1 THE MODEL IN OUTLINE

It is assumed that industry is competitive and distributed into two sectors, unionized and non-unionized (or 'competitive') in a way that is outside firms' control. Firms are able to buy capital goods on an international market at a world real rental cost which is enforced domestically by perfect capital mobility. Each firm enjoys constant returns to scale but is limited by a fixed factor ('entrepreneurship'), so that marginal product declines as the industry expands. It buys imported inputs at a given world price. Accordingly, we write the demand for labour in each sector by profit-maximizing firms as:

$$L_u^d = (\overset{-}{w_u}, \overset{-}{T_F}, \overset{+}{e}, \overset{+}{k}) \tag{5.1}$$

$$L_c^d = (\overset{-}{w_c}, \overset{-}{T_F}, \overset{+}{e}, \overset{+}{k}) \tag{5.2}$$

where u, c subscripts stand for union and competitive sectors respectively, L^d = labour demand, w = real wage, T_F = labour tax rate (as fraction of wage) paid by employer, e = real exchange rate (price of domestic goods relative to price of imported goods, in common currency), k = aggregate (positive) effect of technological progress, real rental on capital, and fixed factor supplies. The expected signs are indicated over the variables.

We complete the description of firms' activities by writing down their production function (we only need the economy's aggregate) as a supply of output equation:

$$Q = (\overset{+}{L_u^d + L_c^d}, \overset{+}{k}) \tag{5.3}$$

where Q = total output of the economy. To avoid aggregation problems we assume the production functions of union and non-unionized industry are identical.[1]

We now turn to the behaviour of workers and unions. Unions maximize the present value of their potential members' aggregated real incomes by setting the union wage. This gives rise to a variable mark-up equation of the form (to be derived below):

$$w_u = m(\overset{+}{\text{UNR}}, \overset{-}{P^{ue}}, \overset{+}{k})w_c \tag{5.4}$$

where m (= one plus the mark up) is a function of UNR (= the unionization rate), k, and P^{ue} (= unanticipated inflation). UNR enters as a proxy for the elasticity of demand for union labour, it being argued that the more unionized an industry, the greater the difficulty of substitution of non-union for union labour in that industry, whether in union firms or by the expansion of non-union firms. P^{ue} enters because unions find it convenient – in order to minimize the

[1] Equation 5.3 is derived from the production function (Equation 5.29) below by substituting for capital (from the marginal productivity condition for capital) in terms of the cost of capital, labour, and the fixed factor.

transactions costs of controlling work conditions – to draw up nominal wage contracts with only partially contingent price clauses; hence a surprise in prices will reduce the real union wage.

It is assumed that firms choose workers' hours given the union-set wage rates. Therefore unionized workers are rationed in their labour supply. We assume that total labour supply of hours in the economy is such that the marginal rate of substitution of leisure for goods equals the *marginal* net real wage available. This is, for union and non-union workers alike, the real wage in the competitive market (which is assumed to be continuously cleared), minus benefits lost and taxes paid through working extra. Because of the wide differences in individual tax/benefit circumstances tight restrictions across the parameters of benefit, tax, and real wage variables are unlikely to hold and we write labour supply, L^s, generally as:

$$L^s = (\overset{+}{w_c}, \overset{-}{b}, \overset{-}{T_L}, \text{POP}) \tag{5.5}$$

where b = real unemployment benefit, T_L = tax rate (fraction of wage) paid by employee, POP = size of (registered) working age population (because this also acts as a proxy for demographic trends, the sign is left ambiguous).

The labour market equations are completed by the equilibrium condition in the competitive sector:

$$L^s - L_u^d = L_c^d \tag{5.6}$$

and by the unemployment relation:

$$U = \text{POP} - L^s \tag{5.7}$$

where U = unemployment. Equation 5.7 states that those registered as potential workers will draw unemployment benefits if not working and will therefore under UK practice be counted as unemployed (this is of course an oversimplification).

Equation 5.6 taken with Equations 5.1, 5.2, 5.4 and 5.5 yields a solution for $w_c, w_u, L_u^d, L_c^d, L^s$, in terms of e, T_F, T_L, UNR, k, and POP. Using Equation 5.3 we can then solve for Q, and from Equation 5.7 for U. The set-up, the economy's 'supply side', is illustrated in Fig. 5.1.

Quadrant (a) shows the equation of the production function (5.3); quadrant (b) shows *SS*, the equation of labour supply (5.5), and *DD*, total labour demand $L_c^d + L_u^a$ in terms of w_c from Equations 5.1, 5.2 and 5.4 the mark-up relation.

We may now conveniently extract from this an open economy supply curve, relating output supply to the real exchange rate, holding the other variables constant. Thus as we lower e, shifting *DD* to the left, we trace out a falling output path along the *PF* curve corresponding to the *DD*/*SS* intersection; this is illustrated by the points (w_{c1}, Q_1, L_1) corresponding to $D'D'$. This is shown in Fig. 5.2, quadrant (a), as *OS*. We trace through in quadrants (b)–(d) the correspondence between real exchange rate, real wages, employment and unemployment, for given other variables. (Note that changes in all these other variables will shift both the *OS* and *EW* curves, and changes in T_L, b, POP will shift the *SS* curve. Also note that the shift in *OS* due to P^{ue} is the 'Phillips curve' effect.)

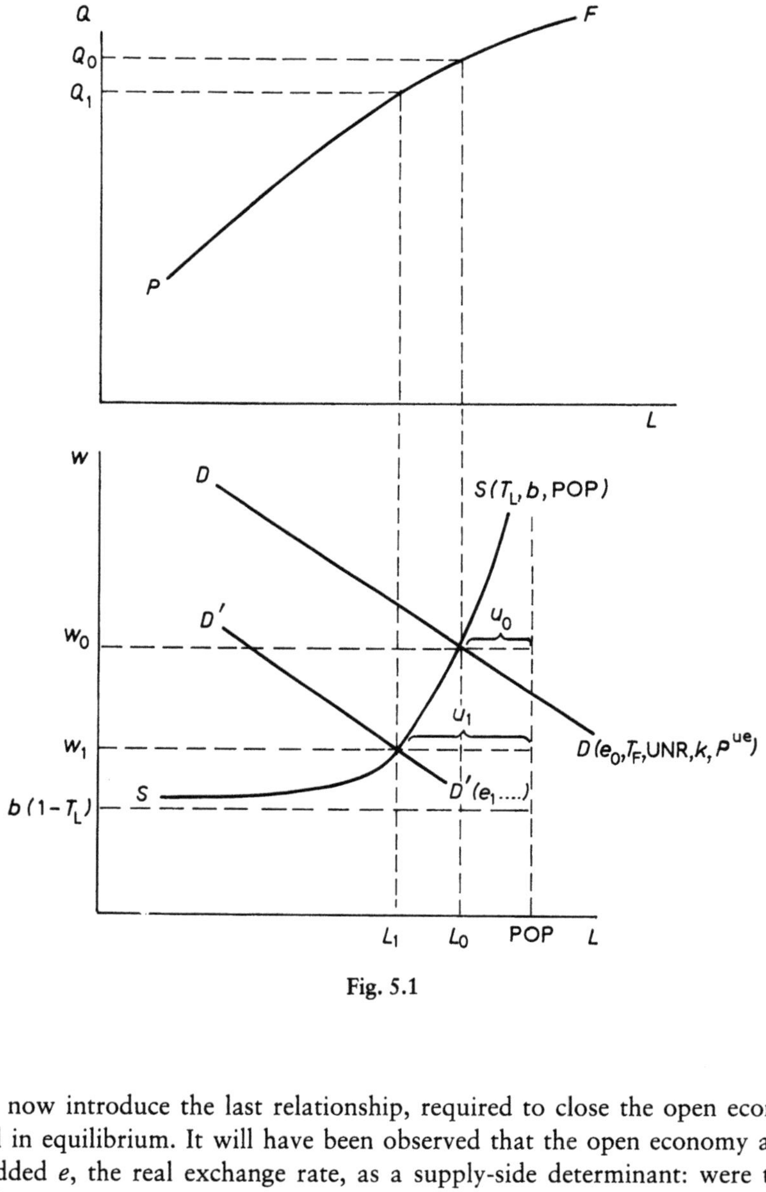

Fig. 5.1

We now introduce the last relationship, required to close the open economy model in equilibrium. It will have been observed that the open economy aspect has added *e*, the real exchange rate, as a supply-side determinant: were this a closed economy, *e* would be absent, and there would be a unique equilibrium supply, corresponding to that which can be produced given labour market equilibrium. This is the usual vertical aggregate supply curve set-up. However, the addition of *e* has produced an upward sloping supply curve; the reason being that as *e* rises, the terms of trade improve and with them profits, enabling firms to induce a higher labour supply profitably.

To close the model we specify a current account balance (x) equation and set

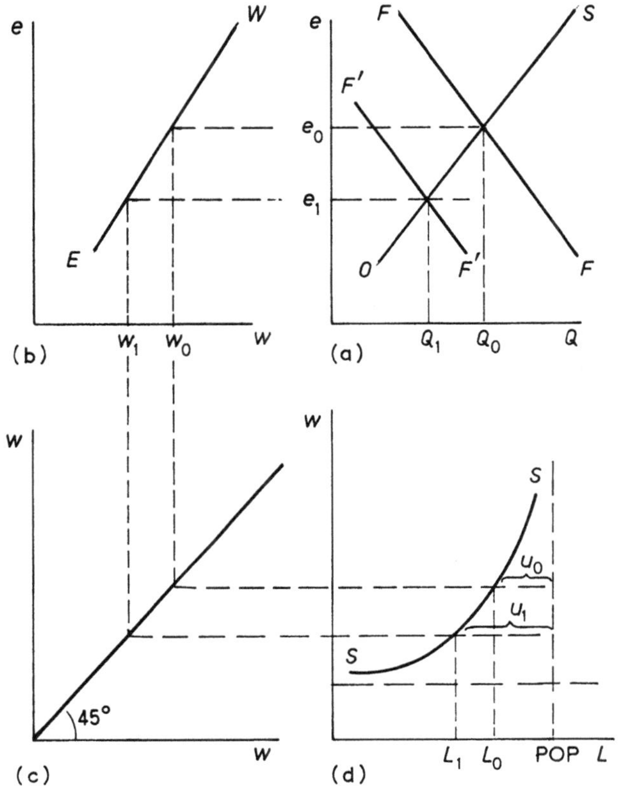

Fig. 5.2

it to equilibrium:

$$0 = x = (\overset{+}{\mathrm{WT}}, \overset{-}{e}, \overset{-}{Q}) \tag{5.8}$$

where WT = the volume of world trade (or output). Equation 5.8 simply states that the demand for imports by UK residents must be equal to the demand for UK exports by foreign residents, equilibrium occurring through *e*, our index of relative home to foreign prices. We can if we wish generalize Equation 5.8 to allow for an equilibrium net transfer (e.g. inwards and a current account deficit for a less developed country, outwards and a surplus for a mature capital exporting country). The *FF* curve in Fig. 5.2, quadrant (a), illustrates Equation 5.8. If this was a 'small' open economy, then the *FF* curve would be horizontal. But in this model this is not an appropriate assumption. Full equilibrium of the economy – with corresponding 'natural rates' – occurs at the intersection of the *FF* and *OS* curves. So one may think of *OS* as describing the short run supply *curve* of the economy, the intersection *FF–OS* as determining the *point* of equilibrium.

To determine the short-run behaviour of the economy requires the addition of an IS, or short-run aggregate demand curve. This we do not specify here as it raises short run macroeconomic issues of impact effects and dynamics that are not our primary concern here. (Full model simulations of these aspects using the 'Liverpool model' (Minford, 1980) of which the supply equations described in this chapter are now a part, are available on request from the author; broad findings about lags are mentioned briefly below.)

In the rest of this chapter we focus in detail on the determination of the *OS* curve by Equations 5.1–5.7. We take Equation 5.8, the *FF* curve, from earlier work (Minford, 1980).

5.2 THE LABOUR MARKET IN DETAIL

5.2.1 Labour market structure

We distinguish a non-union and union sector for a given industry. The non-union sector we suppose to be competitive with atomistic suppliers of labour facing atomistic firms. In Britain, this sector consists primarily of the small business sector of private industry. Examples are the numerous small road haulage firms in the highly unionized (approximately 95%) Transport and Communication sector dominated by British Rail, British Road Services and the National Bus Company; and the large numbers of family engineering firms in the Metal Goods sector (approximately 69% unionized). Additionally the major parts of certain sectors are non-unionized; this applies notably to agriculture (only 22% unionized), construction (27%), and timber and furniture (35%).

The extent of formal unionization is not necessarily a good guide to the extent of union dominance over wages because wage agreements for non-union employees are frequently linked automatically to union agreements (under 'agency shop' arrangements), and on the other hand wage agreements nationally negotiated by unions in highly unionized sectors (such as vehicles or engineering) are merely reference points for shop floor negotiation in individual firms in which the union may be powerless. Mulvey (1976) for example found in a study of 77 Minimum List Heading manufacturing industries in 1976 that the union/non-union wage differential was 0 for industries covered only by national agreements, against 46 to 48% for industries covered by company, district or local government and 41 to 46% for those covered by national with supplementary agreements. Unfortunately coverage data is only available from the early 1970s; and even this is not free of interpretational problems. Unionization is the only reasonably long time-series, and it seems likely to be reasonably well correlated with the extent of union dominance in wage agreements. As a measure of the absolute extent of the non-union sector at a point of time, it is unreliable; but the coverage data confirm the view that there is a significant part of the British economy where unions have no direct influence on wages. For example, in April 1973, as against an average unionization rate of 49%, the extent of coverage by national-with-supplementary or by company, district and local agreements was 43% for manual men, 35% for manual women and much less for non-manual workers. Given that the public sector has very high unionization rates and effective coverage is probably close to 100%, the extent of the non-union sector in Britain's private sector should not be underestimated.

5.2.2 The theory of labour supply

The theory of unemployment that is currently most popular among labour economists is 'search' theory. Suppose that a man has been made redundant. He then searches for a new job for his particular skill. Extra time spent searching costs him extra: this cost includes the outlays on search net of any utility derived from leisure (which of course could be negative). He gets job offers at regular intervals with a wage attached to it which is taken randomly from a distribution (which he knows) of potential wages for the job type. He accepts an offer when it is equal to or greater than the expected wage from the next offer minus the extra cost of search involved in waiting for another period.

This theory is undoubtedly suitable for individuals in certain labour markets: notably, where the individual has a clearly defined job preference and jobs of that type become available periodically, and have a wage distribution attached to them. For example, professional people, such as an economics lecturer, may be well described by it. However, the vast mass of jobs are manual or semi-skilled non-manual; within these jobs some are restricted by union entry conditions, others are in industries with little union intervention. There would seem to be for such jobs a 'going rate', one in the union sector where jobs are rationed, and one in the non-union sector (if there is one for that job type) where jobs are freely available at the rate. Take taxi-drivers for example; there are areas such as Newcastle where there is close regulation of rates and attempted control of entry, and areas such as Liverpool which are effectively deregulated. An unskilled man could become a taxi-driver in Liverpool at will (he simply joins a 'taxi garage', which leases a cab to him); or he could try to get a more profitable regulated job in Newcastle. But it is by no means clear that he will 'search' and remain unemployed. Rather, he may well decide whether it is worth his while to do either job. If he concludes that the deregulated one is good enough, but the regulated one would be better if it came up, he may take the deregulated one and be ready to drop it and shift when and if the other comes up. It seems unlikely that he would remain unemployed, 'searching' the union or regulated sectors, unless he decided that the non-union rates were just not attractive at all; if he did so, he would lose income without necessarily enhancing his chances of a union job.

These considerations suggest an alternative model of the work decision, which is 'new classical' in spirit. The worker has knowledge of 'going rates' in unregulated, or non-union sectors in which he has the necessary skills to work; he does not need to 'search' for this knowledge. He decides when to enter and when to withdraw from these sectors in a standard 'optimizing' manner; i.e. he maximizes the present value of his expected welfare, given these wage rates and other relevant prices, including benefits out of work and taxes etc. in work. Though all workers would like to have a union job, it is assumed that the chances of getting one are not affected by taking a non-union job, so that the union wage does not affect his work decision.

It turns out in this model that the number of people willing to work at *union* wages is irrelevant to the determination of wages or jobs in either the union or the non-union sectors. The reason is straightforward; the union's mark-up over the non-union real wage is determined by its monopoly power interacting with technology and demand conditions; this monopoly power is precisely the power to ignore the desires of non-union members for better wages within the controlled sector. In practice of course this ability would be eroded substantially, the larger the non-union sector; but this erosion depends not so much on the non-union members' frustration as on the enhanced ability of firms to hire non-union members beyond the reach of the union (as in the USA, with firms hiring in the South rather than the unionized North East).

The model therefore implies that the total supply of labour will be dependent on the level of current real wages in the non-union sector ('free market wages'), net of tax and expenses, relative to on the one hand net out of work benefits and on the other expected future net real non-union wages. In other words, the people who are 'on the margin' of supply in the labour market are in the non-union sector (in 'unprotected jobs'), typically on low wages and 'unattractive' jobs: hence the importance in labour supply of replacement ratios for low income households, for these are the ones most likely to withdraw under additional pressure. (By implication, replacement ratios for average workers – who will be unionized typically – are likely to be of no relevance.)

There are various ways in which the supply of labour could contract as real wages fell. Workers could decide to quit more frequently, taking longer periods between work; for manual men for whom explicit part-time work is awkward, this would approximate to part-time working over the year as a whole. Workers could take spells of work abroad, and spells on benefits at home. Workers could decide not to work at all until real wages picked up again: for example, they could withdraw from the labour market in recessions and return in boom periods. Most drastically, they could withdraw indefinitely and change their lifestyle to one of living on benefits and casual, possibly undeclared, earnings, on the assumption that real wages are never likely to be sufficiently attractive. To those accustomed to the ways of prosperous Southern areas of the country, such ideas may seem unfamiliar, even outrageous; but it has to be said that they are part of the everyday gossip and casual empiricism of an area such as Liverpool.

Viewed in this way, the distinctions between decisions on 'duration' of unemployment and those on 'participation in work' become blurred; there is in essence a continuous decision on more or less or no participation in work.

This decision is naturally viewed as taken for the fiscal year. For example, workers wanting to work half the time would be best advised to work half the year and be unemployed for the other half (in either order); that way they use up their tax allowances. Were they to work for one year and be unemployed for the whole of the next year, they would fail to use and subsequently lose their tax allowances in the second year under the UK tax system operating over our sample period.

Let us define π_{it} as the proportion of the current fiscal year's hours the worker i currently plans to devote to leisure on which we assume he obtains a real benefit payment, b_{it}; for his work he obtains the real non-union wage, $w_{ict}(1 - T_{Lt}) = w'_{ict}$ where T_L is the direct tax rate. We assume the worker's adjustment costs are low for non-union employment and can be neglected; this assumption may work rather well for low-paid manual work which constitutes the bulk of non-union employment close to the poverty line where unemployment is heavily concentrated. It also conveniently avoids introducing lags until later in the analysis. We therefore write quite simply that worker i maximizes in each time period his (quadratic) utility.

$$U_{it} = \left[\gamma_i I_{it} - \frac{\beta_i}{2} \pi_{it}^2 + \pi_{it}(\alpha_i + v_{it} + D_t) \right] \tag{5.9}$$

subject to

$$I_{it} = (1 - \pi_{it}) w'_{ict} + \pi_{it} b_{it}; \tag{5.10}$$

I_{it} = his real net income rate; α_i = personal characteristics (e.g. family size) dummies, v_{it} = random factors, and D_t = seasonal factors affecting preferences. The solution from the 1st order condition is:

$$\pi_{it} = \frac{\gamma_i}{\beta_i}(b_{it} - w'_{ict}) + \frac{1}{\beta_i}(\alpha_i + v_{it} + D_t) \tag{5.11}$$

In a cross-section analysis taken at a point in time, π_{it} can be interpreted as the chances of a person being unemployed. In a cohort study of a cross-section over a period of time, π_{it} can be interpreted as the fraction of this whole time period worker i is unemployed.

We could treat all individuals as having the same γ_i, β_i. This would be defensible under the crude quadratic assumption if they all faced similarly sloped budget constraints; for then similar substitution effects in response to benefit changes could be expected. However, in the UK we can distinguish three major groups in the bulk of our sample period, according to their benefit status as illustrated in Fig. 5.3. There are 'flat rate' benefits – i.e. effectively the minimum paid, shown as point A. People on net incomes to the left of A[2] will face a budget line with

[2] The point A is properly the wage at which the flat rate benefit level is just attractive enough to make 100% leisure preferable. This point could occur at a replacement ratio lower than 100% (for some possibly higher).

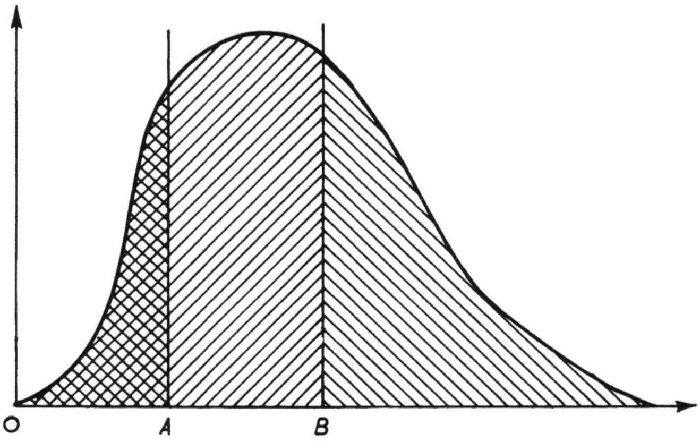

Fig. 5.3 Distribution of income and benefit status for a typical household type. A = flat rate benefit 'cut off' point; B = earnings related benefit 'cut off' point (see text)

no slope. They are likely to be at a corner solution, where they do no work at all, and be unresponsive to benefit or wage changes.

Then there was an earnings related supplement (ERS, abolished December 1981) which raised the benefit/wage ratio to $\frac{2}{3}$ for those whose incomes relative to flat rate benefits put them below this ratio otherwise. Their budget line will have a slope of 33% therefore; they accordingly may be at a tangency point with some substitution effect. The group to the right of A includes these people as well as those with benefit/income ratios between $\frac{2}{3}$ and 1.

Finally, there was a ceiling on ERS, i.e. an income above which ERS became a flat rate supplement, shown as point B. Those to the right of B will have benefit/wage ratios progressively less than $\frac{2}{3}$ as income rises, and their budget lines will slope more steeply; so that their tangency point will be where the substitution effect may be smaller. (At very high incomes budget lines become shallow again with the higher rate of tax; however this is not of much concern for unemployment.)

If we were carrying out a cross-section analysis, it would pay us to use a utility function which spanned these possibilities. However, we are proposing a time-series analysis for which this gives no particular gain. Instead, we note that for our population at t,

$$\pi_t = \sum_{j=2}^{3} q_{jt}\left[\frac{\gamma_j}{\beta_j}(b_{jt} - w'_{jct}) + \frac{1}{\beta_j}(\alpha_j + v_{jt} + D_t)\right] + q_{1t} \qquad (5.12)$$

where q_{jt} = proportion of population in category j (j ranging over the 3 groups just described), and j-subscripted parameters and variables are the averages in each group, q_{1t} is the group to the left of A who are assumed to prefer 100% leisure.

The q_{jt} are themselves functions of the general non-union wage, (w_{ct}), the flat rate benefit, (b_f), the ERS (b_{ers}), and the upper limit (b_{max}), as well as of the income distribution. Most importantly, as flat rate benefits rise, more people will be drawn into the area to the left of *A* where they may decide not to work at all. Since someone who does not work at all will be unemployed 12 times as long as someone who takes a 2 month spell every two years, these people dominate the unemployment stock. In a time-series analysis, we may expect the elasticity of unemployment (under the UK system) to the benefit/income ratio to be very low for a low aggregate ratio (e.g. around 0.5), to rise as the ratio rises, reaching a peak, and dropping again towards zero as the population becomes concentrated around or to the left of *A*. Such a supply curve is illustrated in Fig. 5.4.

This illustrates the important general point that while cross-section studies of populations are undoubtedly of great value, they do not necessarily provide ready answers to questions about reactions of populations over time. In this case to translate cross-section findings into time-series predictions of the effect of benefit changes one requires to know the income distribution, the location of points *A* and *B*, the reservation wage levels, and the elasticity between *A* and *B*. Such a translation is likely to be hazardous at best; a time-series relationship, for all its faults, provides a useful direct estimate of the reactions under investigation.

It is extraordinarily hard to think of tractable set-ups for estimating exactly this time-series model. So far it has eluded us. The discussion must suffice as underpinning for a quite orthodox linear time-series representation. We write unemployment, $U_t = (1 - \pi_t)\text{POP}_t$ as a linear function of benefits, real non-union wages, the direct tax rate, POP (the size of the labour force), and q (a measure of the toughness with which benefit disqualification provisions are enforced).

We set competitive wages as the dependent variable. Later we will do the same when we estimate the labour supply reaction to average wages (i.e. over both

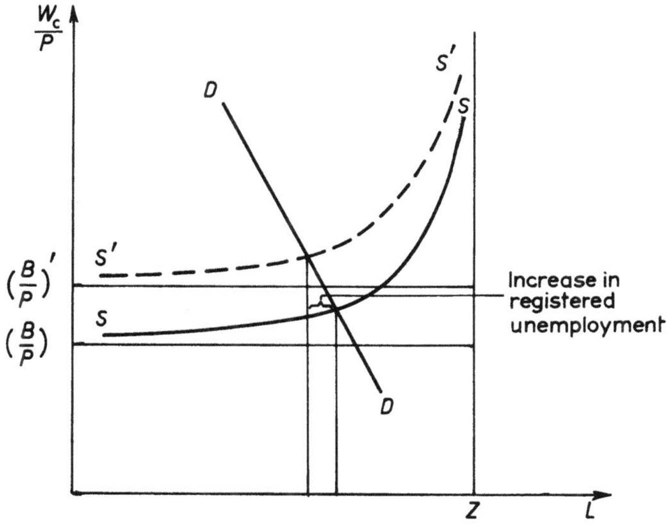

Fig. 5.4 Increase in benefits

union and non-union sectors), on which alone we have data. The decision to normalize on wages is in principle arbitrary and unsubstantive. However, there is a potential practical advantage, which appears to be highly relevant. If the economy has been operating over the sample period in the elastic portion of the supply curve shown in Fig. 5.4, i.e. where benefit wage ratios are rather high and there is a high degree of 'real wage rigidity', α_3 in Equation 5.13 will be rather low, possibly insignificant; normalizing on unemployment would then give poorly determined results, whereas normalizing on real wages will give well-determined ones (this is the same argument as used under conditions of high capital mobility for estimating interest rate rather than capital flows equations). That this is the case is borne out by what we know about benefit/wage ratios for unskilled workers, by the well-known UK facts of real wage rigidity, and indeed by our results which show a low, marginally insignificant, and far from robust unemployment coefficient in the wage equation (also a fairly common finding in standard Phillips curve wage equations). Our equation is therefore:

$$w_{ct} = \alpha_0 + \alpha_1 q_t + \alpha_2 \mathrm{POP}_t + \alpha_3 U_t + \alpha_4 T_{Lt} + \alpha_5 b_t \qquad (5.13)$$

where $b = vb_{ers} + (1 - v)b_f$, v being the weight for those drawing earnings related benefit.

We now turn to the union sector, where we suppose the union to maximize the expected real incomes of its potential members, M (assumed to be exogenous and defined as for example in Bain and Price (1980)):

$$\sum_{i=0}^{\infty} d^i \underset{t-1}{\mathbf{E}} Y_{t+i} \qquad (5.14)$$

subject to

$$Y_{t-i} = w_{ct+i}(M - L_{ut+i}) + w_{ut+i} L_{ut+i} \qquad (5.15)$$

and

$$L_{ut+i} = \beta_0 + \beta_1 k_t - \beta_2 w_{ut} + \lambda L_{ut+i-1} \qquad (5.16)^3$$

[3] Equation 5.16, which is the reduced form demand by firms with rational expectations, is written for simplicity in terms of w_{ut}, k_t only, on the assumption that these follow a first order autoregressive process. Generalization to higher order and moving average processes complicates Equation 5.21 but does not alter its nature, i.e. planned real union wages remain a function of lagged (possibly more than once) real wages and of expected future competitive real wages and technology. β_2 also depends on the union's preferences because these influence the process driving w_{ut}. However, allowing for this also makes no essential difference to Equation 5.21, as can be seen by rewriting Equation 5.16 as

$$L_{ut+i} = \beta_0 + \mathbf{E} \sum_{t+ij=0}^{\infty} \beta_3^i (\beta_1 k_{t+i+j} - \beta_2 w_{ut+i+j}) + \lambda L_{ut+i-1},$$

where β_3 is the inverse of the unstable root in *firms'* Euler equation, treating w_{ut} as exogenous. The equivalent of Equation 5.20, for large i, then becomes

$$\underset{t-1}{\mathbf{E}}\left[\left\{1 - \left(2\frac{(1+\lambda\beta_3)}{\beta_3 + d\lambda}\right)B + \left(\frac{\lambda + d^{-1}\beta_3}{\beta_3 + d\lambda}\right)B^2\right\} w_{ut+i+1} = \frac{-1}{\beta_3 + d\lambda}[(1 - \beta_3 B^{-1})(1 - \lambda B)W_{ct+i} \right.$$

$$\left. + \beta_2^{-1}(1 - d\lambda B^{-i})(1 - d^{-1}\beta_3 B)\beta_1 k_{t-i} + \beta_2^{-1}\beta_0(1 - \beta_3)(1 - d\lambda)(1 - d^{-1}\beta_3)]\right]$$

Factorization of { } yields the equivalent of Equation 5.21 for w_{ut+i} (large i); w_{ut+i} tends towards this process, because the importance of β_3 is less than smaller i.

The latter demand for labour function is the linear analogue of Equation 5.30 below (but with lagged adjustment); k_t represents the effects on labour productivity of capital costs, technological progress and other factors, L_{ut} = labour demand in union firms, w_{ut} = union real wages, $0 < d < 1$ is the union discount factor. (The real exchange rate, e_t, is substituted out in terms of that part which responds directly to w_{ut}, included in β_2, and the rest, included in β_1.)

This is a standard treatment in which members are assumed to be risk-neutral and those who lose their jobs are treated as having the same marginal utility of income as those who obtain higher wages; but see Oswald (1982) for a discussion of the theoretical underpinnings of the union decision, including this set up forced on us here for practical reasons.

The first order conditions yield:

$$\underset{t-1}{\mathrm{E}}\left[w_{ut+i} - w_{ct+i} = \frac{L_{ut+i}}{\beta_2} - \frac{d\lambda(w_{ut+i+1} - w_{ct+i+1})}{1 - d\lambda B^{-1}}\right] \tag{5.17}$$

or

$$\underset{t-1}{\mathrm{E}}\,[w_{ut+i} - w_{ct+i} = (1 - d\lambda B^{-1})\beta_2^{-1}L_{ut+i})] \quad i \geqslant 0 \tag{5.18}$$

where B is the lag operator instructing us to lag variables but *not* the expectations operator, and the $\underset{t-1}{\mathrm{E}}$ outside the square brackets indicates that all variables are expected values conditional on information at $t - 1$.

Hence using Equation 5.17:

$$\underset{t-1}{\mathrm{E}}\left[w_{ut+i} - w_{ct+i} = \beta_2^{-1}\frac{(1 - d\lambda B^{-1})}{(1 - \lambda B)}(\beta_0 + \beta_1 k_{t+i} - \beta_2 w_{ut+i})\right] \quad i \geqslant 0 \tag{5.19}$$

so that:

$$\underset{t-1}{\mathrm{E}}\left\{\left[1 - \frac{2}{d\lambda}B + \frac{1}{d}B^2\right]w_{ut+i+1}\right.$$
$$\left. = -\frac{1}{d\lambda}(1 - \lambda B)w_{ct+i} - \frac{\beta_1}{\beta_2 d\lambda}(1 - d\lambda B^{-1})k_{t+i} - \frac{\beta_0(1 - d\lambda)}{\beta_2 d\lambda}\right\} \quad (i \geqslant 0) \tag{5.20}$$

The transversality condition for the union is that the rate of increase of real wages as given by Equation 5.20 converges in infinite time to the constant rate determined by the trends of w_{ct} and k_t; any other solution is suboptimal (in fact 'a road to ruin' leading to the extinction of the union either by pricing its members totally out of the union industry market or by pricing them so far below the non-union industry market as to cause them all to leave the union). This condition is met, following the operator method of Sargent (1979, e.g. pp. 333–8) by factoring

$$\left(1 - \frac{2}{d\lambda}B + \frac{1}{d}B^2\right) \quad \text{as} \quad (1 - \rho_1 B)\left(\frac{1 - (\rho_2 B)^{-1}}{-(\rho_2 B)^{-1}}\right)$$

where ρ_1 is the stable root and ρ_2, the unstable root, is used forwards. For a unique stable solution we require that one root is stable, the other unstable (i.e.

there is a saddlepath solution), which is satisfied here since

$$\rho_2 = (d\lambda)^{-1}[1+\sqrt{(1-d\lambda^2)}] > 1. \quad \rho_1 = (d\rho_2)^{-1} = \frac{\lambda}{1+\sqrt{(1-d\lambda^2)}} < 1.$$

The solution for w_{ut} is then:

$$\underset{t-1}{\mathbf{E}}\, w_{ut} = \rho_1 w_{ut-1} + \frac{1}{\rho_2 d\lambda}\sum_{i=0}^{\infty} \rho_2^{-i} \underset{t-1}{\mathbf{E}}\Big[(1-\lambda B)w_{ct+i} + \frac{\beta_1}{\beta_2}(1-d\lambda B^{-1})k_{t+i} + \frac{\beta_0}{\beta_2}(1-d\lambda)\Big] \quad (5.21)$$

In other words, the planned real union wage depends upon initial levels and the expected path of non-union real wages and other cost factors. We shall rewrite Equation 5.21, using the F^* prefix to denote the summed and discounted forward values of w_{ct} and k.

We also note that β_2 will fall as expected union power increases, measured by UNR (the unionization rate). The larger the union sector as a proportion of any given industry, the greater the difficulty of substituting non-union for union labour (a) by unionized firms (directly), (b) by the non-unionized firms (indirectly) through their own expansion. This difficulty is due to 'primary' strike action against (a) and union firms, which would be breaching implicit or explicit 'closed shop' arrangements, and 'secondary' strike action against (b) and non-union firms, because they would be interfering with unions' industrial purposes. Thus at any given real union wage, higher union density in an industry (and so greater strike power) lowers the elasticity of demand for union labour; this is true also in the long run, because the strike threat operates against free entry into the non-union sector. For these reasons, using an expansion of the terms in k and the constant, we add an argument in UNR to Equation 5.21. So we write (dropping hereinafter the t-subscripts throughout):

$$\underset{-1}{\mathbf{E}}\, w_u = v_0 + v_1 F_w^* w_c + v_2 F_k^* k + v_3 F_u^* \mathrm{UNR} + \rho_1 w_{u-1} \quad (5.22)$$

We wish to incorporate the observed facts of nominal wage contracting by a proportion of unions. The typical settlement length in our sample period has been one year, and we suppose that the proportion with 1 year contracts is c, the rest implicitly recontracting continuously via shop floor alterations of piece-rates, overtime etc.

For the latter, w_u is given by Equation 5.22; they will succeed in setting w_u at the planned level. The former however set *nominal* wages, W_u, consistently with Equation 5.22, given the information set at the settlement date and the expected price level over the coming year so that:

$$W_u = 0.25\left(\underset{-i}{\mathbf{E}} \sum_{j=0}^{3} P_+\right) + \underset{-1}{\mathbf{E}}\, w_u \quad (5.23)$$

Given that a constant proportion settle each quarter, the change in contracted nominal wages in any quarter will reflect (apart from planned real change) 0.25 times the average inflation rate expected at $t-1$ for the coming year (PEXP_{-1}). Real wages however will change by this deflated by the actual change in prices since the last quarter. Hence actual real wages will be approximately:

$$w_{\mathrm{u}} = \frac{W_{\mathrm{u}}}{P} \approx \underset{-1}{\mathbf{E}}\, w_{\mathrm{u}} - (\Delta \log P - 0.25\,\mathrm{PEXP}_{-1}) \tag{5.24}$$

for values of $\underset{-1}{\mathbf{E}} w_{\mathrm{u}}$ close to a (normalized) unity.

Putting contracts and non-contracts together yields:

$$w_{\mathrm{u}} = v_0 - cP^{\mathrm{ue}} + v_1 F^*_{\mathrm{w}} w_{\mathrm{c}} + v_2 F^*_{\mathrm{k}} k + v_3 F^*_{\mathrm{u}} \mathrm{UNR} + w_{\mathrm{u},-1} \tag{5.25}$$

where $P^{\mathrm{ue}} = (\Delta \log P - 0.25\mathrm{PEXP}_{-1})$ is 'unanticipated inflation'.

Our (only) observable series for wages is the total:

$$w = uw_{\mathrm{u}} + (1-u)w_{\mathrm{c}} \tag{5.26}$$

Substituting from Equation 5.25 we get:

$$\begin{aligned} w = {} & uv_0 - ucP^{\mathrm{ue}} + uv_2 F^*_{\mathrm{k}} k + uv_3 F^*_{\mathrm{u}} \mathrm{UNR} \\ & + \{uv_1 F^*_{\mathrm{w}} + (1-u)(1-\rho_1 L)\} w_{\mathrm{c}} + \rho_1 w_{-1} \end{aligned} \tag{5.27}$$

Referring back to Equation 5.13 we may observe that the arguments of w_{c} consist of tax and benefit rates, demographic trends, benefit disqualifications, and the unemployment rate. In Equations 5.13 and 5.22 it is natural to assume – and is not broadly violated by the data – that the tax and benefit rates follow a random walk, and that the demographic trends, UNR, q, and k are random walks with a time trend. This allows us to transform these variables into their current and lagged values, constants, and time trends. This leaves the expected unemployment rate and we shall write $\{uv_1 F^*_{\mathrm{w}} + (1-u)(1-\rho_1 L)\}U$ as a function of the lagged and current unemployment rate and of the *natural* rate, U^*, at t; the latter at this stage we have been forced to proxy by a time trend.

Finally we obtain our wage equation as:

$$\begin{aligned} \log w = {} & c_0 - ucP^{\mathrm{ue}} + c_1 q + c_2 \mathrm{UNR} + c_3 \log \mathrm{POP} + c_4 T_{\mathrm{L}} \\ & + c_5 \log b + c_6 \log U + \rho_1 \log w_{-1} \end{aligned} \tag{5.28}$$

We have adopted a loglinear formulation and have dropped lagged values for certain variables. Forced to choose between a time trend and log POP because of their high collinearity, we omitted the (generally quite insignificant) time trend. Nothing serious appears to hang on these approximations.

We choose to enter log U rather than U into this supply curve because our theory suggests that, at high unemployment levels, a 1% change in benefits will have a larger absolute effect on unemployment than at low unemployment levels because the slope of the supply curve will be flatter (more 'wage rigidity'). A log formulation has this property.

The relation between Equation 5.28, our 'supply' equation, and the equilibrium model of Lucas and Rapping (1969) can be seen by rewriting Equation 5.28 as:

$$U = (\overset{-}{P^{ue}};\ \text{UNR},\ \overset{+}{\text{POP}},\ T_L,\ b,\ w_{-1};\ \bar{w}) \qquad (5.28')$$

(28)′ indicates that unanticipated inflation, P^{ue}, reduces unemployment. The role of 'expected normal real wage' is taken by the variables between semi-colons; these depress labour supply (raise registered unemployment). Finally current real wages raise labour supply and depress unemployment. Of course the derivation and true meaning of Equation 5.28′ is quite different from that in the Lucas equilibrium model. But the family resemblance is striking; were someone to use the Lucas model on an 'as if' basis for the UK, he should obtain reasonable results (e.g. Batchelor and Sherriff, 1980).

Equation 5.28 also has a family relationship with the wage equation suggested by Hines (1964), who first stressed the role of union power in wage-setting, in our view quite correctly. Nevertheless, as might be expected from the twenty year gap in formulations, ours is different in important ways. The most important one is that the Hines wage equation was intended as a 'cost–push' theory of *nominal* wages and so of inflation; this formulation was effectively rebutted by Purdy and Zis (1973). Equation 5.28, however, has no implications for inflation, other than temporary, and relates to *real* wages.

5.3 THE DEMAND FOR LABOUR

Individual firms are regarded as competitive and atomistic each with a fixed endowment of entrepreneurship which gives them decreasing returns to scale. They therefore determine an optimal supply of output and demand for labour given the actual and expected prices and wages facing them.

We write the *i*th firm's production function in the form

$$Q_i(1-\mu) = f(K_i, L_i, \bar{T}_i) \qquad (5.29)$$

where μ is the (assumed inflexible) share of imports in production and $\bar{T}$ is the stock of the exogenous factor ('entrepreneurship') which is assumed to be growing steadily over time. Using the cost-minimizing conditions and the production function, we can write down a labour demand function in terms of output as:

$$L_i^d = \phi(w_i(1+T_F),\ r_i,\ Q_i,\ \bar{T}_i) \qquad (5.30)$$

We can also substitute in this for output, from the profit-maximizing condition that marginal cost = price, and so obtain the reduced form demand function for labour as:

$$L = \psi(w_i(1+T_F),\ e,\ r_i,\ \bar{T}_i) \qquad (5.31)$$

which corresponds to Equations 5.1 and 5.2 above; *e* is the relative price of imported inputs (the real exchange rate as before), r_i the real cost of capital.

It is convenient to base estimation on Equation 5.30 treating *Q* as planned (or 'desired') output; Equation 5.33 below is its form in implementation. We then use an estimate of the marginal cost curve below, Equation 5.38 (taken from Minford (1980)) to complete the supply model; from Equations 5.33, 5.38 and the labour supply Equation 5.28 we then derive the 'OS' curve relating e to Q in Fig. 5.2, explained above.

Writing Equation 5.29 more generally, also in an adjustment formulation and aggregating over the i firms gives familiarly:

$$\Delta \log L^{D} = \mu(\log L^{*} - \log L_{-1}) \tag{5.32}$$

where $L^{*} = \phi[(1 + \bar{T}_{F})\bar{w}^{*}, \overset{+}{Q}^{*}, t]$ and t picks up the trend in $\bar{T}$ and r.

This formulation, though *ad hoc*, corresponds approximately to the final form of an optimal plan by individual firms (e.g. see Nickell, 1981). The starred variables are the desired or expected future outcomes, as appropriate. Since we use unemployment as the quantity measure, our operational demand equation becomes:

$$\log U = \mu\gamma_1(\log w^{*} + T_{F}) + \mu\gamma_2 \log Q^{*} - \mu\gamma_3 t + (1 - \mu)\log U_{-1} \tag{5.33}$$

(Note that any residual effect of POP is captured here by the time trend.) We continue to use log U for convenience given its use in Equation 5.28; we test later whether this choice causes empirical difficulties.

Equations 5.28 and 5.33 define a sub-model of the labour market which is conditional on planned output. Planned output depends upon the rest of the model interacting with this sub-model. The sub-model aggregates over *all* sectors of the economy, including nationalized industries and public services. We therefore treat a rise in unit labour costs in the public sector (due say to increased public sector unionization) in the same way as one in the private sector. The reason is that the government budget constraint compels the government to spend the resources generated by given levels of taxation and borrowing; *given* these levels a rise in public sector labour costs forces substitution and contraction effects on labour demand just as in the private sector.

It is true that there is an aggregation problem since these levels will be changing over time and, because labour intensity may differ between public and private sectors, so changing labour demand. We treat any such drift as exogenously determined by political choice (on tax burdens and, for borrowing, on inflation targets); but it is included in the 'productivity' time trend along with other sources of trend such as the efficiency of taxation, relative raw material costs and real capital costs, none of which we could in practice identify separately. (This applies also to rising unionization which would in principle lower γ_1 over time as discussed above, and contribute a compensating time-dependent alteration to the constant term.)

5.3.1 Empirical work on UK labour market

Our labour demand and supply equations have been deliberately kept to the simplest form in order to economize on degrees of freedom. Possibly intensive searching would yield some insights into nuances of dynamic structure; but we doubt if much could be gained, because the major exogenous variables in the labour subsector are benefit and tax variables, for which the best predictor turns out to be close to the current value. The only exogenous variable to show significant autocorrelation is UNR; this both displays a strong upward time trend and positive first order autocorrelation which could form the basis for a better predictor. Planned/expected output creates more of a problem; Nickell (1981) reports encouraging results for a labour demand function in which quite complicated patterns of coefficients on future expected values are generated. At this stage, we have merely looked at expected values up to a year ahead using instrumental variables for the quarterly version and the model predictions for the annual.

Our study relates to most of the post-war time series data conveniently available from the Central Statistical Office's data bank; namely quarterly data for the whole economy, for manufacturing and for individual industries and for annual data for the whole economy only. The Liverpool model is annual; but the other work serves as a further check and is possibly more informative because quarterly.

5.3.2 Whole economy time series: quarterly data

The results that follow are based on benefits and average net direct taxes paid by a married man with two children whose wife is not working. Clearly this is not properly representative of those on the margin of employment and unemployment. We have also therefore checked the sensitivity of our results for a variety of weighting procedures; a full set of results for benefits and tax series weighted according to unemployment shares for six household-types (single, married +0 up to +4 children)is available on request but is not different in any important way from those shown here.

The disqualification variable turned out to be generally insignificant in the quarterly analysis (e.g. Table 5.5). An annual series was also not available for our full annual sample. We therefore dropped this variable altogether from the main reported results.

Our quarterly results for the whole economy are shown in Table 5.3. All variables have the correct sign and are significant. In the 'supply' or 'wage' equation we tested the following restrictions: that the coefficient on benefits (b) and tax rates (T_L) were (a) equal to each other and (b) also both equal to unity in the long run. (a) would come about if $\alpha_4 = \alpha_5 = 1$ in (13), i.e. if the benefit/net non-union wage *ratio* affected the supply of labour, an assumption that is widely made in the literature . (b) would occur if additionally in Equation 5.27 $\{uv_1F_w^* + (1-u)(1-\rho_1 L)\}w_c \approx (1-\rho_1)w_c$, a condition that in general seems unlikely to be met. In general we find that (a) is met but (b) is not and report the results for (a) accordingly.

The fact that benefits and direct taxes have an equal coefficient in the supply or wage equation, but one that is significantly less than unity in the long run, suggests that versions of this equation with unemployment as the dependent variable and the benefit/income *ratio* as a regressor are mis-specified (unless the level of real wages is also included); the reason as already discussed lies in the way w_c enters the supply equation which is as a complex dynamic term.

In so far as one is interested in the long-run partial elasticity of unemployment to the replacement ratio, this equation provides an estimate of $4\frac{1}{2}$, well above previous time series of the UK. Maki and Spindler (1975) found a partial elasticity for the postwar period of 0.6. This estimate is flawed by the use of the ERS benefit ratio which only applied to a minority of unemployed, and by the ambiguous status of the estimating equation, which is neither a structural equation nor a reduced form. Also, the most recent re-working of their data by Junankar (1981) reveals that the equation is fairly vulnerable to shifts in estimation period. Holden and Peel (1981) have estimated reduced form equations on UK postwar time series using benefits paid to a married man on average earnings with 2 children which find an elasticity of around 0.4 to the replacement ratio; however since other exogenous variables, such as world trade and unionization, have been omitted and rolled into the error process, the coefficient estimate may be biased. Similar problems and comments arise in the context of the work on inter-war data by Benjamin and Kochin (1979) and others (see Benjamin and Kochin *et al.* 1982). This type of time-series work has been stimulating and suggestive; but it would appear that it has established only that benefits probably matter for unemployment, to an *undetermined* degree.

Unionization enters the supply equation strongly. The long-run elasticity of real wages to UNR (at its mean value of 0.48) is about unity.

It is tempting to attempt to derive an estimate of the union mark-up from this coefficient. In so far as any tentative estimate of this union mark-up emerges from our work, it is of the order of 60%.[4] Other estimates have varied substantially, and have given rise to substantial controversy. Most recently, Treble (1982) has pointed out the tenuous basis for the majority of these estimates which

[4] An estimate of the *increase* in the union mark-up due to increased unionization can be obtained from the model as follows. Reduced form effects give

$$\frac{d\log U}{d\mathrm{UNR}} = 11.5; \quad \frac{d\log w}{d\mathrm{UNR}} = 0.8; \quad \frac{d\log U}{d\log w_c} = -10.0$$

(partial elasticity from supply equation); $d\log w = \alpha d\log w_u + (1-\alpha)d\log w_c$ where α is the proportion of workers covered by union agreements (according to New Earnings Survey 1978 this was 0.71 for men and 0.68 for women, if *all* agreements are included). Hence

$$\frac{d\log w_u - d\log w_c}{d\mathrm{UNR}} = \frac{1}{\alpha}\left(\frac{d\log w}{d\mathrm{UNR}} - \frac{d\log w_c}{d\mathrm{UNR}}\right) = \frac{1}{0.7}\left(0.8 + \frac{11.5}{10.0}\right) = 2.79$$

The change in UNR over our sample period 1955–79 was 0.14, hence the change in log (w_u/w_c) is put at 0.39 implying a rise in the mark-up of 48 percentage points. Assuming that the mark-up was quite low in 1955 (say 10% as in some early studies such as Pencavel (1974) on 1964 data), we could take 60% as an appropriate estimate with 50% as lower bound. This is subject to the qualification that *other* factors than UNR could have affected the mark-up.

use a methodology originated by Lewis (1963). If one averages available estimates of this type for the UK mark-up (Parsley, 1980), it comes out around 25%, with a high variance. Treble (1982), used a random coefficients estimator to rework the data of Mulvey (1976), who gave estimates from the UK around 30%, and obtained an estimate of about 40%. This is closer to our implicit figure; however there is still reason to believe it is downward biased.[5]

Another issue on which these results throw some light is the prevalence of one-year non-contingent contracts. The term P^{ue}, will have a coefficient of minus unity if such contracts are universal in the sector under investigation; its absolute size below unity indicates the proportion of workers governed by such contracts. It is normally regarded as a 'stylized fact' about the UK that most wages are governed by such contracts. Indeed this is confirmed by evidence from wage *rates*. Yet in Italy where inflation rates have been similar over the past decade, indexation is virtually universal, and this of course is equivalent to a shortening of contract length. Belgium and the Netherlands also have widespread indexation in spite of a lower inflation experience. It is also the case that recorded labour contract lengths (on wage rates) declined in the UK from about 2 years in the early 1950s to around 1 year by the mid-1970s (Minford and Hilliard, 1978). There is therefore some evidence both that UK contract behaviour is sensitive to the inflation rate and yet that it is out of line with some continental behaviour.

Our results which relate to earnings suggest that one year contracts are far from universal. In fact the coefficient on P^{ue} was ultimately dropped in the aggregate economy equation because it had the wrong sign, besides being insignificant. It

[5] Treble's criticism and proposal can be simply reproduced. Lewis proposed using the identity

$$\log W_i = \log W_{Ni} + \mu_i \alpha_i$$

where W_i = average wage. W_{Ni} = non-union wage, μ_i = union mark-up, α_i = proportion unionized (all in industry i at a point in time): letting $\mu_i = \mu$ and $\log W_{Ni} = f(\chi_i) + \varepsilon_i$ where χ_i is a vector of industry characteristics, he estimates

$$\log W_i = f(\chi_i) + \mu\alpha_i + \varepsilon_i$$

Treble points out that μ is not constant and that the χ_i have been chosen in *ad hoc* manner; consequently he argues that estimates of μ have varied widely because of omitted variable bias (χ_i) and because the error contains the error in μ, θ_i, in the form $\theta_i\alpha_i$ and so is correlated with α_i. He proposes that since the χ_i are arbitrary, they be dropped and that μ be written $\delta + \theta_i$, and the equation be estimated by the random coefficients method as

$$\log W_i = \delta\alpha_i - (\varepsilon_i + \theta_i\alpha_i).$$

The problems however go deeper than Treble suggests. Clearly micro theory would lead us to expect $\log W_{Ni}$ to depend negatively, and μ_i to depend positively, on α_i: it also should suggest an appropriate list of χ_i apart from α_i. Thus the relation should be:

$$\log W_i = f(\chi_i, \alpha_i) - \mu(\alpha_i)\bar{\alpha}_i - \varepsilon_i$$

Therefore in addition to the problems Treble mentions there is the bias from omitting α_i from the model for $\log W_{Ni}$. The coefficient on α_i estimated by Mulvey *et al.* includes the negative effect on $\log W_{Ni}$ and is hence downward biased. This downward bias of course persists in Treble's own reworking. Furthermore, Treble's removal of all χ_i from the regression increases the likelihood of omitted variable bias.

is possible to reconcile this finding with the prevalence of 1 year contracts for wage *rates* by appealing to wage drift. Presumably matters are arranged on the shop floor so that earnings are indexed to some degree to the inflation rate. Variation in overtime and piece rates or clocking on/off conventions, could well be used to achieve this effect. On this basis the UK contract behaviour is seen to be roughly in line with the continental practices cited. At the same time there seems no reason to believe that the *change* in contract length evidenced by wage rate data would be seriously misleading (indeed, Minford and Hilliard found it to reflect the predictions obtained from a transactions–cost minimization approach).

Other features of the real wage equation are the (on the whole) significant role of unemployment, the fairly speedy adjustment rate (90% in 2 years), and the modest negative effect on real wages of the rising workforce. The unemployment term implies that the supply curve of labour is not infinitely elastic ('real wage rigidity') over the *whole* sample period; the (partial) long run elasticity of unemployment to real wages is around $-9\frac{1}{2}$, probably implying (Table 5.1) an elasticity of *labour supply* to real wages of over unity and to real benefits of about -0.5. Nevertheless, prior to 1973 the unemployment term is insignificant (Table 5.4) indicating a much higher elasticity (and so greater real wage rigidity).

The negative effect of the rise in workforce presumably reflects its bias towards married women and juveniles, who have it would seem generated some lowering of the real supply price of labour.

The speed of adjustment should reflect approximately that of unionised firms in their labour demand decisions (i.e. in the U equation). No cross-equation constraint was applied but the free estimates are encouragingly close.

The demand for labour (here the unemployment) equation shows strong effects of output, real wages grossed up for employee taxes paid by employers, and productivity. The restriction that real wages and taxes have the same coefficient is rejected; this odd result does however not occur for manufacturing (results not reported) and suggests it may be due to unusual behaviour in the non-manufacturing sector (especially perhaps the public sector). The unrestricted equation (Table 5.5) gives a much higher coefficient on taxes than on wages which is hard to rationalize (and requires more research, probably on expectational aspects). We therefore prefer and have reported the restricted estimate.

We may use the quarterly results to estimate the partial elasticity of supply and demand for labour approximately, if we assume some relation $\partial \log L/\partial \log U$. Supposing that changes in employment are exactly equal, though oppositely signed, to changes in unemployment, then

$$\frac{\partial \log L}{\partial \log U}=\frac{\partial L/L}{\partial U/U}=\frac{-\partial U/L}{\partial U/U}=\frac{-U}{L}=-\text{(unemployment rate)}$$

However, owing to failure to register by a proportion (for fear of disqualification, partly), this will be an underestimate. If we impose $\partial \log L/\partial \log Q=1$ (assuming that fixed entrepreneurship does not introduce significant diseconomies of scale

Table 5.1 Elasticities

Demand for labour	To real wages (plus N.I. contributions)	−0.5
	Output	1.0 (imposed)
Supply of labour	To real wages	1.2
	Real benefits (grossed up for employees' taxes)	−0.6

within the region of our sample) in our demand equation, for the rest we obtain the elasticities given in Table 5.1. (The implied $\partial \log L/\partial \log U = -0.118$ implying $\partial L = -2.4\partial U$ at the mean unemployment rate of 0.05. This implication that the unemployment rate would rise by about 1 percentage point for every $2\frac{1}{2}$% drop in output, *cet par.*, is roughly consistent with Okun's law.)

Various issues can be raised with these equations.

As far as specification tests are concerned, the equations are reasonably invariant both to changes in time period and to relatively obvious specification changes. Tables 5.4 and 5.5 illustrate. Stability over time periods is fairly impressive; both equations pass the Chow test and an *F*-test for equality of coefficients.[6] The key parameter values do not move much; the only exception is the real wage cost coefficient in the unemployment equation, which drops away if experience from the mid-sixties is not included. The reason for this is that much of the rise in T_F, employers' taxes, occurred between 1964 and 1967 (Fig. 5.8); excluding this data therefore both removes interesting variation and alters the trend in this variable.

The unemployment equation is free of autocorrelation, as indicated by the X_4^2 test. The wage equation suffers from 3rd order autocorrelation; the X_4^2 test is just significant at the 5% level, and it turns out that the residual at the 3rd lag is also just significant in the regression for that test. However, the autocorrelation is modest; correction for it makes no difference to the point estimates, though it makes them better determined (Table 5.5, column 1).

Inclusion of a time trend in the wage equation and the labour force in the unemployment equation leaves parameter values unchanged; and both additional variables are insignificant.

Various other variables have been suggested as potential candidates for inclusion in the wage equation, within its overall specification. These include: the age structure of the population (those less than 20, A20, those more than 55, A55), the degree of toughness in denying benefit (TOUGH), inflows (INF) and outflows (OUT) from the unemployment register. Table 5.5 shows results with a variety of these. They are uniformly disappointing however. Table 5.5 also shows that,

[6] The Chow test give *F*-values for the wage equation of 1.1 (truncated at 76.4), 0.6 (truncated at 67.1), for the unemployment equation of 1.04 and 1.0 respectively. The *F*-tests for constancy of coefficients (with the sample split in 1973) were 0.68 for real wages and 0.11 for unemployment.

if benefits and direct tax are split up, the coefficients are not significantly different, though that on tax is less well determined and the higher of the two.

Our choice of log U rather than U in the demand function was made for convenience and could perhaps cause a loss of definition in the estimates. Table 5.5 shows (last column), however, that it does not. Significance levels are broadly unchanged as are the elasticity estimates at the mean; there is a slight rise in the coefficient on output, a slight fall in that on labour costs. We therefore maintained the log form.

Is there reverse causation from real wages to real benefits and if so is it such as to bias our results? The vast proportion of benefits paid out (i.e. excluding the earnings related supplement which applied to under 20% of the unemployed) have been determined by reference to flat rate entitlements and not by a ratio to earnings, and these entitlements are fixed about 6 months in advance of the year to which they apply. Furthermore, the entitlements have been mainly governed by political considerations (e.g. the Wilson government's decision to raise benefits substantially in the late 60s) and the ratio to real wages has been permitted to vary considerably, as real wages have varied; this suggests that politicians have had in their minds varying 'poverty base lines' which have borne little systematic relationship since the mid 60s with real wages. It is however possible that current price changes affect both real wages and real benefits jointly. We used a version of the Wu test of exogeneity (Smith, 1983) to check this; this regresses real wages on an Instrumental Variable (I.V.) estimate of real benefits, the other variables, and the residual of benefits from this I.V. estimate. Exogeneity is tested by the significance of this residual. As Table 5.6 shows, there is some evidence of endogeneity, but it does not bias the coefficients seriously as the equation with the I.V. estimate of benefits above reveals.

To test more specifically whether the common price deflator itself was generating the relationship between real wages and real benefits, we included it (on an I.V. basis) in the regression (Table 5.6, column 5). It can be seen that it is entirely insignificant.

We also found (Table 5.6) no significant relationship from *lagged* real wages to real benefits, which suggests that whatever is determining benefits it is probably not real wages. In fact, real benefits follow a random walk according to this work. This is not too surprising a result with such a highly politicized variable.

Unionization itself may respond to real wages and unemployment, and hence may bias the coefficients. We also carried out the Smith/Wu test for this, with a clearly negative result (Table 5.6).

It may be asked: how far can the equilibrium relationships Equations 5.28 and 5.33 estimated here be distinguished from disequilibrium relationships of a standard form? In the first place, our Equation 5.33 is regarded as reflecting the demand for labour; it is also common in disequilibrium models to assume that employment is 'on the demand curve' (e.g. because wages are set by noncontingent contracts where firms choose quantity), and if so to use equations amounting to Equation 5.33.

The distinction lies in Equation 5.28, a 'supply price' equation, as compared with the 'sticky wage change' equations used in disequilibrium models. A multiplicity of these 'Phillips curves' exist in the literature; these take the form:

$$\Delta \log W = f(\Delta \log P^{E}_{-1}, U, \log w^{*}, \log w_{-1}) \qquad (5.34)$$

where $\Delta \log P^{E}$ = expected inflation, log w^{*} = equilibrium real wage. If (a) the arguments of log w^{*} are chosen in the same way as by us, (b) rational expectations are used for $\Delta \log P^{E}$ and (c) a unit coefficient were imposed on it then indeed the resulting equation will be indistinguishable from (5.28).[7] But by the same token there would be little dispute about the causes and natural rate of unemployment!

This manifestly is not the case because the Phillips curve form typically used in disequilibrium models has none of the properties (a)–(c). Such a typical Phillips curve in UK models is that due originally to Sargan (1964) and since widely used (e.g. in the NIESR model, Brookes and Henry (1983)). It is of interest therefore to compare this form with Equation 5.28.

We therefore carried out a non-nested test. We estimated a Sargan-type (S) equation over our quarterly data, obtaining a prediction series log $\hat{W}_s$. To obtain a prediction series, log $\hat{W}_{(5.28)}$, for our equation we took the predictions for log w from Equation 5.28 and added to them our I.V. estimate of log P. (S) and Equation 5.28 were then reestimated with respectively log $\hat{W}_{5.28}$ and log W_s as additional arguments; the significance of these indicates the extent to which the alternative model rejects the equation being estimated. The results (Table 5.7) show that log $\hat{W}_s$ is totally insignificant in Equation 5.28, while log $\hat{W}_{5.28}$ has a t-value of 1.5 in (S), just short of significance. While Equation 5.28 cannot quite reject S, S certainly cannot reject Equation 5.28.

The test is handicapped by the failure to compare like with like; 5.28 is a *real* wage equation and to compare it with S, a nominal equation, log P has to be added, in principle from the Liverpool model as a whole but in practice here from an I.V. estimate. The properties of the test with I.V. estimates as arguments are also not known. Nevertheless, there is certainly no prima facie evidence here against Equation 5.28.

This conclusion is strengthened by the following consideration: the Sargan equation can be regarded as an unrestricted multivariate time series representation of our model. Fitted to the same data, such a representation is therefore likely to be reasonably good, and any rejection by a properly restricted model could be expected to be marginal on a fairly small sample. (For issues raised in this type of comparison, see Wallis, 1980).

5.3.3 Annual data

A word is necessary about the annual results for 1955–79, Table 5.2. These were estimated using *expected* output and inflation from the Liverpool model. As in

[7] Our model, as an equilibrium model, implies that unemployment is voluntary whereas such a disequilibrium model would imply involuntary unemployment. For this reason it is unlikely that the arguments of w^{*} would be the same, even if (b) and (c) were adopted.

the quarterly equations the contract (unanticipated inflation) term is either zero or the wrong sign; nevertheless, this is surprising because the data are for wage *rates* where contracts are unambiguously important. This may be the result of aggregation over the calendar year. However, the long run coefficients are similar to the quarterly ones for 1964–79. Such differences as there are (i.e. rather higher coefficients on UNR and POP) are presumably the result of the rather longer time period covered.

Empirically, therefore, the whole economy equations give a reasonably robust performance. Nevertheless, they are based on a limited number of observations and one would wish to corroborate the coefficient estimates from other evidence.

5.3.4 Other evidence

The model we described earlier can be used at the industry level as a 'micro' model. In our unemployment registration system people register for particular jobs so that they can be classified by industry; we interpret the unemployment figures by industry as mirroring the original activities of the unemployed and so as closely reflecting changes in employment in that industry. Besides unemployment, we have real wages and output by industry, and unionization rates in 13 of the industries; where we do not have them, we have used some closely matching rates and failing that the aggregate unionization rate. The other variables do not vary across industries and remain our aggregate figures.

Our full industry results are set out in Minford *et al.* (1983, pp. 129–30). Table 5.8 (row 5) gives a summary. In all industries bar one, the benefit variable has a highly significant effect on real wages, while unionization in the industry is significant in eight of the industries and has *t*-values over 1.5 in another two. In the unemployment equations, in 15 of the industries real wage costs are significant, often highly so, and in the other two they have the appropriate sign and are merely not so well determined.

Hence for our key coefficients, for benefits, union power, and wage costs, there is widespread evidence in the industry level data. This represents strong and striking confirmation of our model of wages and unemployment at the most detailed level; there are over 1000 observations in these industry samples.

Further evidence on these relationships is available both from regional UK data for 1978 and on time series for Belgium, whose system of benefits and degree of union power show close parallels with that of the UK. Analogous relationships with variations of this model reflecting institutional differences in each country have also been detected for Germany, France, and Italy – see Minford *et al.* (1983).

SOURCES AND DEFINITIONS FOR TABLES 5.2–5.7

U	Unemployment	*D of E Gazette*
Q	Total output of all industries	CSO
w	Real average earnings	CSO

Table 5.2 3SLS estimates of supply and demand equations on annual data (t-values)

1955–1979	$\bar{R}^2$	*DW*	*BP(4)*	*S.e.e*
$\log U_t = 40.72 + 1.89(\log \hat{w} + T_F)_t - 2.63 \log Q_t^{EXP}$	0.87	1.51	13.1	0.184
(2.03) (1.10) (1.30)				
$+ 0.058t + 0.52 \log U_{t-1}$				
(0.84) (2.85)				
$\log w_t = 6.26 - 0.062 \log \hat{U}_t + 0.12(\log b + T_L)_t$	0.96	2.51	6.3	0.026
(1.16) (1.53) (2.90)				
$+ 0.92\text{UNR}_t - 0.70 \log \text{POP}_t + 0.80 \log w_{t-1}$				
(2.66) (1.43) (4.72)				

Unexpected inflation failed to enter with the right sign. w = wage rates.
Q_t^{EXP} = prediction from Liverpool model of next year's output, based on this year's exogenous variables (and last year's endogenous).
Instruments used for variables with a circumflex^: T_{Ft}, T_{Lt}, Q_t^{EXP}, $\log U_{t-1}$, t, $\log b_t$, UNR_t, $\log \text{POP}_t$, $\log w_{t-1}$.

Table 5.3 Real wage and unemployment equations for whole economy

(t-values bracketed) *Quarterly: 1964(2) to 1979(2)*

$$\log w_t = \underset{(1.5)}{0.34} - \underset{(2.75)}{0.00162\text{D1}} + \underset{(0.41)}{0.0024\text{D2}} + \underset{(0.77)}{0.0044\text{D3}} - \underset{(2.1)}{0.024 \log \hat{U}_t} + \underset{(3.1)}{0.11\{1 \log b) + T_{Lt}\}}$$

$$+ \underset{(2.4)}{0.45\text{UNR}_t} - \underset{(0.1)}{0.016 \log \text{POP}_t} + \underset{(12.0)}{0.77 \log w_{t-1}}$$

$\bar{R}^2 = 0.47$; S.e.e. $= 0.0125$; $\chi_4^2 = 10.3$*; DW $= 1.719$

$$\log U_t = \underset{(5.2)}{13.3} + \underset{(0.01)}{0.000419\text{D1}} - \underset{(3.52)}{0.125\text{D2}} + \underset{(0.13)}{0.0047\text{D3}} - \underset{(4.5)}{2.17 \log \hat{Q}_t} + \underset{(2.7)}{1.04\{(\log \hat{w}_t) + T_{Ft}\}}$$

$$+ \underset{(2.6)}{0.008t} - \underset{(11.3)}{0.744 \log U_{t-1}}$$

$\bar{R}^2 = 0.96$; S.e.e. $= 0.4986$; $\chi_4^2 = 2.9$*; DW $= 2.008$

Instrumental estimates shown by a circumflex (^).
Instruments used: (Lagged) inflation, output, real wages, real benefits, unemployment; (current) change in log £M3. T_F, UNR, T_L, D1, D2, D3.
* This is the Lagrange Multiplier test for residual independence (the regression residuals are regressed on the right hand side variables and their lagged values, up to 4 in this case; $T \times R^2$ where T is the sample size is distributed as χ^2). The critical value of χ_4^2 is 9.5 at the 5% level (11(1) at the $2\frac{1}{2}$% level). See Godfrey (1978).

Table 5.4. Real wage and unemployment equations for different samples (whole economy) (constant and seasonal dummies included but not reported)

Dependent variable	$\log w_t$					$\log U_t$					
Sample	*64(2)–79(2)*	*64(2)–73(1)*	*73(2)–79(2)*	*64(2)–76(4)*	*67(1)–79(2)*	*Sample*	*64(2)–79(2)*	*64(2)–73(1)*	*73(2)–79(2)*	*64(2)–76(4)*	*67(1)—79(2)*
$\log \hat{U}_t$	0.024	0.010	0.033	0.022	0.027	$\log \hat{Q}_t$	2.17	3.36	2.2	2.48	2.2
	(2.1)	(0.5)	(1.6)	(1.8)	(1.7)		(4.5)	(3.4)	(3.3)	(4.7)	(4.6)
$(\log b_t) + T_{Lt}$	0.12	0.06	0.126	0.13	0.11	$(\log \hat{w}_t) + T_{Ft}$	1.04	1.69	0.07	1.7	0.18
	(3.2)	(0.7)	(1.4)	(2.9)	(2.0)		(2.7)	(1.3)	(0.1)	(3.2)	(0.3)
UNR_t	0.45	0.63	0.48	0.65	0.47	t	0.008	0.010	0.013	0.005	0.013
	(2.4)	(1.8)	(0.9)	(2.3)	(1.8)		(2.6)	(0.7)	(2.2)	(1.3)	(3.4)
$\log \text{POP}_t$	0.02	0.21	0.09	0.2	0.006	$\log U_{t-1}$	0.74	0.78	0.71	0.70	0.68
	(0.07)	(0.4)	(0.1)	(0.7)	(0.02)		(11.3)	(8.7)	(7.0)	(9.5)	(9.0)
$\log w_{t-1}$	0.77	0.72	0.77	0.68	0.78						
	(12.2)	(3.7)	(7.4)	(6.3)	(11.0)						
$\bar{R}^2$	0.98	0.95	0.86	0.98	0.97		0.96	0.94	0.90	0.94	0.94
SEE	0.0155	0.0103	0.0183	0.0152	0.0161		0.097	0.062	0.116	0.096	0.097
RSS	0.01242	0.00181	0.00875	0.00975	0.010912		0.4987	0.0693	0.3624	0.3985	0.4035
DW	1.72	1.95	1.62	1.74	1.67		2.0	1.56	2.16	2.12	2.05

Instruments as in Table 5.3.

Table 5.5. Some alternative specifications. Quarterly 1964(2) to 1972(2) (constants and seasonal dummies included but not reported)

Dependent variable	$\log w_t$								
					1964(2) to 1978(4)				64.2 to 77.4
$\hat{p}_t^{ue}$	0.58 (1.3)	0.3 (0.4)	0.29 (0.5)	0.4 (0.7)	—	—	—	—	—
$\log \hat{U}_t$	−0.030 (1.8)	−0.023 (1.7)	—	−0.023 (1.8)	−0.029 (1.7)	−0.02 (1.6)	−0.02 (1.3)	−0.02 (2.0)	−0.02 (1.7)
$\hat{U}_t$ (millions)	—	—	−0.028 (1.8)	—	—	—	—	—	—
$\log b_t$	—	0.11 (1.9)	—	—	—	—	—	—	—
$(\log b_t) + T_{Lt}$	0.14 (4.5)	—	0.125 (3.4)	0.11 (1.9)	0.12 (3.1)	0.13 (3.3)	0.14 (2.6)	0.12 (3.2)	0.13 (2.7)
T_{Lt}	—	0.16 (1.0)	—	—	—	—	—	—	—
UNR_t	0.48 (5.6)	0.44 (2.3)	0.50 (2.6)	0.40 (1.5)	0.47 (2.2)	0.43 (2.2)	0.37 (1.5)	0.44 (2.2)	0.35 (1.6)
t	—	—	—	0.003 (0.2)	—	—	—	—	—
$\log \mathrm{POP}_t$	−0.004 (0.8)	−0.002 (0.01)	0.035 (0.2)	−0.05 (0.2)	0.04 (0.3)	−0.08 (0.3)	−0.07 (0.3)	−0.02 (0.1)	−0.2 (0.6)
$\log w_{t-1}$	0.75 (14.1)	0.76 (11.1)	0.73 (10.8)	0.74 (8.6)	0.78 (12.1)	0.76 (11.3)	0.76 (11.6)	0.77 (11.8)	0.76 (8.2)
log (INF)	—	—	—	—	0.01 (0.5)	—	—	—	—
log (OUT)	—	—	—	—	—	−0.02 (0.8)	—	—	—
TOUGH	—	—	—	—	—	—	—	—	−0.007 (0.6)
A55	—	—	—	—	—	—	—	0.0001 (0.5)	—
A20	—	—	—	—	—	—	−0.04 (0.5)	—	—
ρ_{-3}	−0.35	—	—	—	—	—	—	—	—

Dependent variable	$\log U_t$		U_t (million)
$\log \hat{Q}_t$	−2.43 (5.6)	−2.13 (4.3)	−2.02 (6.1)
$\log \hat{w}_t$	−0.04 (0.1)	—	—
$(\log + \hat{w}_t) + T_{Ft}$	—	1.23 (2.5)	0.56 (1.8)
T_{Ft}	2.0 (4.7)	—	—
t	0.019 (4.6)	0.006 (1.3)	0.009 (3.6)
$\log U_{t-1}$	0.59 (8.1)	0.75 (11.2)	—
U_{t-1}	—	—	0.79 (12.8)
$\log N_t$	—	0.78 (0.5)	—

Table 5.6 Tests of exogeneity for b and UNR. Quarterly: 1964(2) to 1979(2) (constants and seasonal dummies included but not reported)

Dependent variable			$\log w_t$				$\log b_t$
$\hat{p}^{uc}$	0.42	−0.04	0.14	—	—	$\log b_{t-1}$	0.96
	(0.7)	(0.1)	(0.3)				(18.4)
$\log \hat{U}_t$	−0.022	−0.023	−0.023	−0.020	−0.018	$\log U_{t-1}$	0.009
	(1.8)	(1.8)	(2.0)	(1.8)	(1.4)		(0.5)
$(\log \hat{b}_t) + T_{Lt}$	0.12	0.09	—	—	—	$\hat{p}^{ue}_t$	−0.15
	(3.0)	(2.2)					(0.3)
$(\log b_t) + T_{Lt}$	—	—	0.12	0.11	0.13	$\log w_{t-1}$	−0.06
			(3.1)	(2.8)	(3.5)		(0.5)
UNR_t	0.45	0.49	—	—	0.59	T_{Ft}	0.05
	(2.2)	(2.4)			(1.7)		(0.4)
$\log \mathrm{POP}_t$	0.005	−0.16	0.03	−0.03	0.19	t	−0.0001
	(0.02)	(0.7)	(0.1)	(0.1)	(0.4)		(0.1)
$\log w_{t-1}$	0.76	0.84	0.77	0.79	0.76	$\bar{R}^2$	0.93
	(11.1)	(14.2)	(11.4)	(12.4)	(11.9)		
$\log b_t - \log \hat{b}_t$	0.13	—	—	—	—	DW	1.76
	(1.85)						
$\widehat{\mathrm{UNR}}_t$	—	—	0.42	0.40	—	Box—Pierce	11.0
			(2.0)	(1.9)		(χ^2_{16})	
$\mathrm{UNR}_t - \widehat{\mathrm{UNR}}_t$	—	—	0.63	—	—		
			(1.4)				
$\log \hat{P}$	—	—	—	—	−0.28		
					(0.6)		

Instruments as in Table 5.3.

T_F	Employers' NI contributions as percentage of gross (national) average earnings, male manual workers national figures.	*Annual Abstract of Statistics* (D of E)
UNR	National unionization rate	*D of E Gazette*
POP	National working population	*Economic Trends*
P_{ue}	$\Delta \log P - (\mathrm{PEXP}_{-1}/4)$	
PEXP_{-1}	Expected average change of price level in coming year (national expectations in the sense of Muth, 1961), i.e. $$\log[E_{-1}(P_0 + P_1 + P_2 + P_4)/4] - \log(P_{-1} + P_{-2} + P_{-3} + P_{-4})/4$$ The expected value in this expression $E_{-1}(\ )$ was derived by an instrumental regression on lagged values and on rates of change of output and money supply.	

Table 5.7 Tests of wage equation ((5.28) in nominal terms) against Sargan equation (*S*)

	S equation		*Quarterly 1964.2 to 1979.2* *(5.28) = Wage equation (nominal)*	
Dependent variable	$\log W_t$		$\log W_t$	
$\log W_{t-1}$	1.0	1.0	$\log \hat{P}_t$	1.0
C	0.25	0.32	C	1.14
	(2.1)	(2.5)		(0.3)
$\Delta \log P_{t-1}$	0.33	0.38	$\hat{P}^{ue}$	0.22
	(2.1)	(2.4)		(0.6)
$\log U_{t-1}$	−0.019	−0.020	$\log \hat{U}_t$	−0.015
	(1.9)	(1.9)		(1.4)
$\log W_{t-1} - \log P_{t-1}$	0.001	−0.041	$(\log b_t) + T_{Lt}$	0.11
	(0.02)	(0.6)		(3.4)
t	0.0008	0.0019	UNR_t	0.42
	(1.5)	(2.1)		(1.4)
$\log \hat{W}_{t(5.28)}$	—	−0.31	$\log \mathrm{POP}_t$	−0.10
		(1.5)		(0.3)
D1	−0.001	−0.002	$\log W_{t-1} - \log P_{t-1}$	0.78
	(0.2)	(0.3)		(11.0)
D2	0.012	0.012	$\log \hat{W}_{t(S)}$	−0.004
	(2.3)	(2.3)		(0.1)
D3	−0.010	−0.011	D1	−0.005
	(1.8)	(2.0)		(0.9)
SEE	0.01443	0.01427	D2	−0.001
DW	1.58	1.55		(0.1)
			D3	0.004
				(0.8)

Instruments as in Table 5.3. $\log \hat{W}_{t(5.28)}$ is prediction of (5.28) + $\log \hat{P}_t$. $\log \hat{W}_{t(S)}$ is prediction of *S* equation.

b	A weighted average of real benefits (national figures) $\left[\frac{(B + \mathrm{ERS})0.3 + (B + R + M)0.7}{P}\right]$	
B	Flat rate benefits, man with wife and second child	DHSS
ERS	Earnings related supplement based on gross average earnings	Holden and Peel (1979)
R	Local authority rents	D of E
M	School meals	DES
T_L	Percentage of national gross average earnings lost in taxes and national insurance contributions (national figures) (man with wife and second child)	Holden and Peel (1979)

P	Retail price index (all items)	*Economic Trends*
	denotes use of instrumental variable estimate	

5.3.5 The natural rate – estimates and policy effects

An idea of how sensitive the long run unemployment effects are to the different data examined here can be obtained from Table 5.8.

By comparison, Nickell (1979a, b), Lancaster (1979) and Mackay and Reid (1972) have found elasticities of unemployment duration to the replacement ratio of around 0.6 in cross-section studies of samples of the unemployed. Lancaster went as far as to conclude that 'an elasticity of this order could now be regarded as established beyond reasonable doubt'. These studies employ a totally different methodology and data set from that used here; until our methodology has been tried out properly on such cross-section data, it would be inappropriate to comment on these estimates in detail.

Nevertheless, there is one major issue to be raised with all these studies. They all assume, within a search model framework, that intended (or desired) duration is never long term (or 'infinite'), or in terms of the search model that the offer-acceptance rate never tends to zero. We have suggested above that a person's optimal level of unemployment per fiscal year will depend sensitively on the slope of his budget constraint (as roughly measured by the replacement ratio). For high ratios (very shallow slopes) he may go to a 'corner solution' and decide to work not at all, or only for brief spells when market wages are exceptionally favourable. Such people are likely to exhibit a very low elasticity to changes in the ratio; this is Nickell's finding for those on long durations (six months or more). Yet the implication, if our suggestion is correct, is precisely the opposite to that which he draws (that there is little effect of the ratio on long term unemployment); it is that at some ratio these people would cease to have a corner solution and would participate 'properly' again in employment, having therefore at this ratio a very high ('switching') elasticity to it.

Table 5.8

% change of unemployment with respect to*	*% change in benefits*	*Percentage point changes in:* Unionization rate	*Personal tax rates*	*Employer tax rates*
From aggregate model				
(Annual data)	1.5	11.3	1.5	2.5
(Quarterly data)	2.8	11.5	2.8	5.9
From average of industry models	4.0	9.3	4.0	8.4
Liverpool model	2.8	11.5	2.8	5.9

* *Not* percentage point change. Calculations assume that each set of equations for w and U is combined with rest of Liverpool model (i.e. Equations 5.37 and 5.38).

We now proceed to evaluate the properties of the equations for the whole economy. These are based on the quarterly estimates of the long run coefficients, being the better-determined, except for the productivity growth time trend which comes from the annual unemployment equation (fitted over the full sample period, it appears more reliable). The lagged dependent variable coefficients are taken from the annual equations. The long run coefficients of Equations 5.28 and 5.33 in the Liverpool model are accordingly:

$$\log U^* = -8.48(\log Q^* - 0.022\text{time[years]}) + 4.06(\log w^* + T_F) \quad (5.35)$$

$$\log w^* = -0.10 \log U^* + 0.48(\log b + T_L) - 0.07 \log \text{POP} + 1.96\text{UNR} \quad (5.36)$$

The coefficients of $\log U_{t-1}$ and $\log w_{t-1}$ are respectively 0.52 and 0.80.

In the rest of the model the long-run relationship between real wages and output, when absorption is constrained by the condition that the current account is in balance is:

$$\log Q^* = -0.65e^* + 0.46 \log \text{WT} \quad (5.37)$$

and the long-run relationship between the real exchange rate and real wages (given by the pricing equation) is:

$$e^* = 1.86(\log w^* + T_F) - 0.033t \text{ (years)} \quad (5.38)$$

where WT = index of world trade volume, e = real exchange rate.

We can use these four equilibrium equations to compute the effects on the natural rate of unemployment, real wages and output of various permanent changes in taxation, benefits and unionization – see Table 5.9. (All tax and benefit changes are offset notionally by lump sum transfers, leaving net government revenues unchanged; these, which include the effect of changes in output, are noted in column 5).

It can be seen that per unit of revenue cost cuts in national insurance (N.I.)

Table 5.9 Effects of regime changes[3]

Fall of	*Unemployment*[1] *('000)*	*Real (%)*[1] *wages*	*Output*[1] *(%)*	*Available*[2] *for lump sum transfers (£billion p.a., 1982 prices)*
10% in benefits	−500	−2.1	+2.5	+3.4
0.01 in T_F	−90	−0.6	+0.5	−0.6
0.01 in T_L	−55	−0.2	+0.25	−0.9
0.01 in UNR	−170	−0.8	+1.0	+1.1

[1] Computed on 1980 values.
[2] Includes extra revenue from rise in output. Negative figure denotes net drain on Exchequer. Assumes marginal overall tax rate of 0.4.
[3] *Source*: Simulations of Liverpool model.

charges paid by firms are more effective in reducing unemployment than cuts in the standard rate while to equal the effect on unemployment of a 10% cut in real benefits it would require cuts in N.I. charges costing the Exchequer a net £6 billion p.a. more. However, the effectiveness of a cut in taxes on employees in reducing unemployment would presumably be increased the more it was concentrated on the lower paid; one may presume that if totally concentrated on those at the bottom end each £ of tax cut would have an effect on unemployment comparable with that of a £ of benefit cut. At present most of those in the 'unemployment trap' would pay significant tax if they were working (Minford *et al.*, 1983).

The lags before these policy effects come through depend upon the dynamics of the policy variables. For benefits, unionization, and income taxes one third of the full effect comes in year 1, $\frac{2}{3}$ by year 2, 80% by year 3, and all by year 5. For employers' taxes the lags are much shorter (since the direct effects on labour costs are large and immediate): $\frac{2}{3}$ of the full effect comes in year 1, and all by year 2.

Finally, we may use these equations to compute the natural rate in 1980. Figures 5.5–5.10 that follow show the model's 'prediction' for this from 1956–80, as well as the behaviour of the exogenous variables. The natural rate of unemployment is estimated at about $1\frac{3}{4}$ million ($7\frac{1}{4}$%) in 1980; roughly this figure also emerges from both quarterly and annual wage/unemployment equations run on their own over their respective samples treating actual output exogenously at y^*. The date at which it began its dizzying rise can be set fairly precisely at 1965, i.e. the beginning of Labour government after 13 years of Tory government. This led to a sharp rise in union power, in benefits and in taxation.

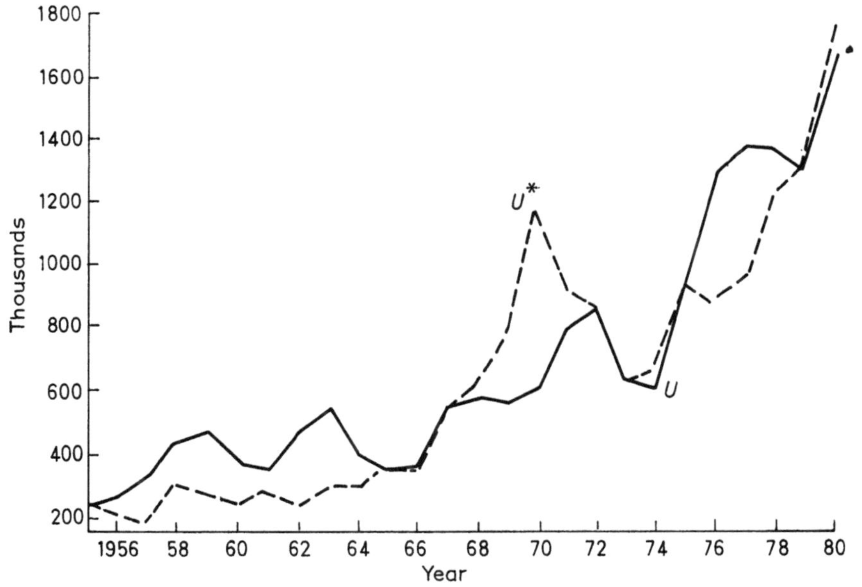

Fig. 5.5 *U* (unemployment) and *U** (equilibrium unemployment)

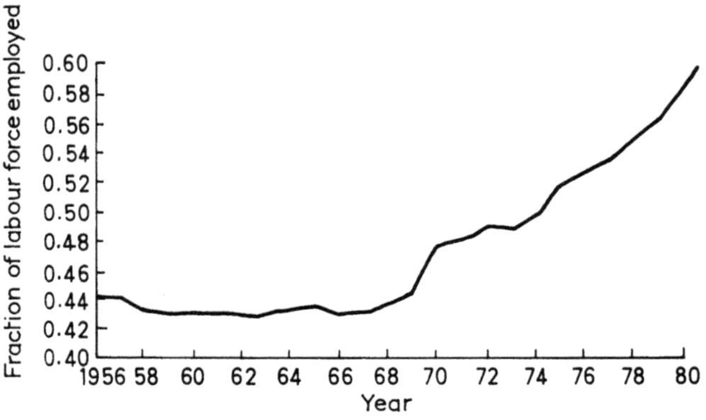

Fig. 5.6 UNR (unionization rate)

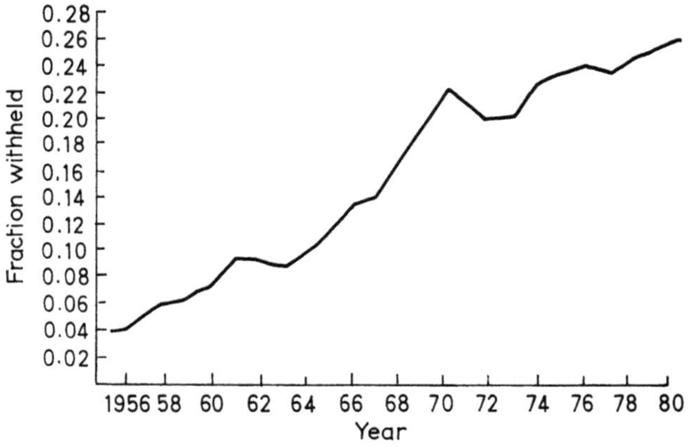

Fig. 5.7 T_L (direct tax rate)

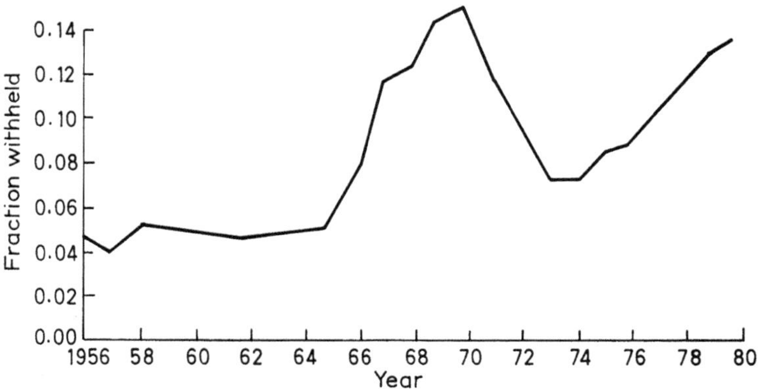

Fig. 5.8 T_F (employers' tax rate on labour)

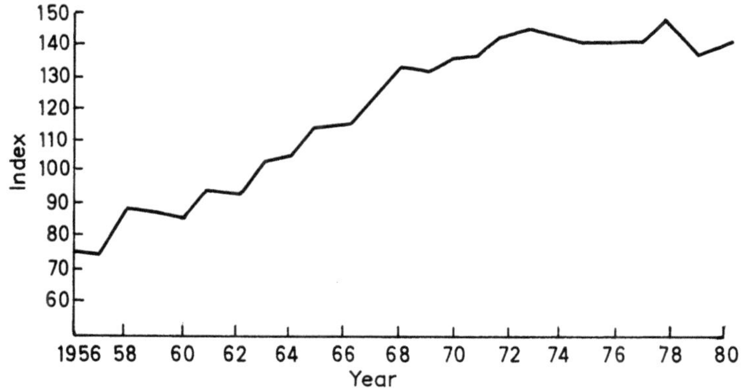

Fig. 5.9 *b* (real benefits)

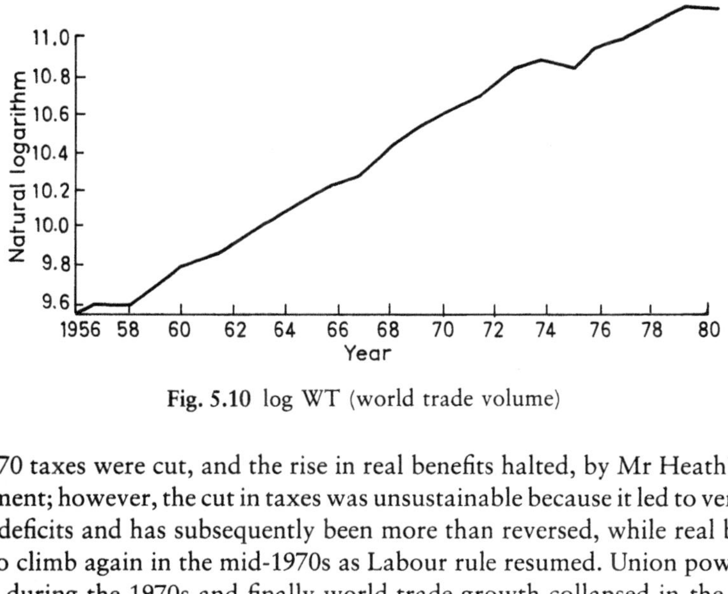

Fig. 5.10 log WT (world trade volume)

In 1970 taxes were cut, and the rise in real benefits halted, by Mr Heath's Tory government; however, the cut in taxes was unsustainable because it led to very large budget deficits and has subsequently been more than reversed, while real benefits began to climb again in the mid-1970s as Labour rule resumed. Union power rose steadily during the 1970s and finally world trade growth collapsed in the second half of the 1970s. So the upward trend was resumed from 1973.

5.4 RESPONSES TO CRITICS

Since this work was originally published in Minford (1983) there have been a number of criticisms of it and the related ideas on policy in Minford *et al.* (1983). Apart from policy issues which are ignored here, these criticisms relate to general and technical matters.

The main general criticism from Keynesians has been that 'demand is not taken into account', that unemployment is a problem of deficient demand, and that forces acting on supply are irrelevant because labour has been in excess supply and involuntarily unemployed. Thus if supply-side measures caused more people to wish to work (reducing 'voluntary' unemployment), there would be no reduction in total unemployment, no rise in employment, merely a shuffling of the unemployed from the voluntary to the involuntary category. In this model it is assumed that there is a residual 'non-union' market in which any excess supplies of labour in the economy are eliminated; this 'clears' the labour market. In this sense, there is never any involuntary unemployment even in the short term.

This is a powerful assumption but it cannot be checked for correctness by direct observation, as some critics have implied. You cannot look at the labour market and see whether it is in chronic excess supply or in continuous balance. This assumption can only be tested indirectly by checking the predictive accuracy of models embodying it ('equilibrium' models) against others 'disequilibrium' models). Some evidence is offered by the success of this model relative to disequilibrium one, as reported above.

However, it is no part of my thesis – or of any equilibrium analysis for that matter – that demand 'does not matter'. On the contrary, unemployment is caused, in the short and medium term, by shocks to demand (such as deflationary policies). How much unemployment at any one point in time can be ascribed to this source is an important question which my paper indirectly addressed by asking how much is due to underlying or long-term factors. In short, the point made is that these factors, mainly non-demand factors, *do* matter.

I turn now to technical criticisms, that is, those concerning the variables chosen to determine wages and unemployment and the soundness of the empirical results. These have been dealt with in detail in two journal replies (Minford, 1984, 1985), so what follows is a non-technical summary.

Professor Steve Nickell of Oxford University (Nickell, 1984) has criticized the treatment of productivity in this model. He argues that in the wage equation there should be an explicit variable to proxy (expected) productivity since this influences firms' ability to pay. To this I have no objection, having included a number of proxies which turned out to contribute not at all to the explanation. However, his particular suggestions for proxies, the capital/labour force ratio and the price of imports relative to that of British goods, are objectionable because they are both endogenous variables (that is, they respond to the same causal pressures as do wages, employment, output, etc.). Their inclusion in the wage equation will induce a bias particularly in annual data where simultaneity is more important than in quarterly data. Indeed, when included in annual data they do seriously distort the results, eliminating the impact of benefits and 'attributing' much of the rise in real wages to a rising capital stock (a pretty suspect result when one considers that rising wages are likely themselves to lead to a rising capital/labour ratio). However, this equation exhibits bad statistical properties (the errors in it are very badly 'autocorrelated' or non-random). Furthermore, when included in quarterly data – on which the results are the most clear and reliable, there being much more of it in the sample — Nickell's variables make no significant difference.

Dr Brian Henry, and two colleagues (Henry *et al.*, 1985) of NIESR have argued that in the quarterly wage equation, on which they rightly focus in preference to the annual one, the proxy for union power, the unionization rate, is insignificant if the sample is extended to 1982. They argue from this that there is no evidence for a union-induced effect on real wages. But, first, we would certainly expect the

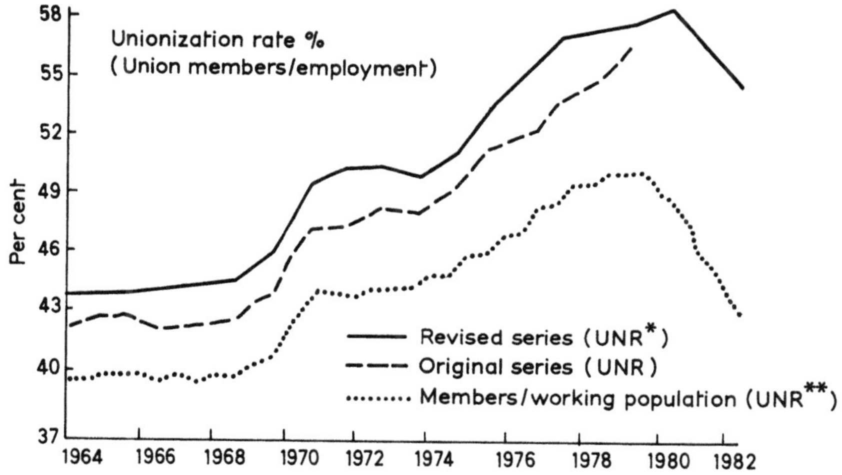

Fig. 5.11

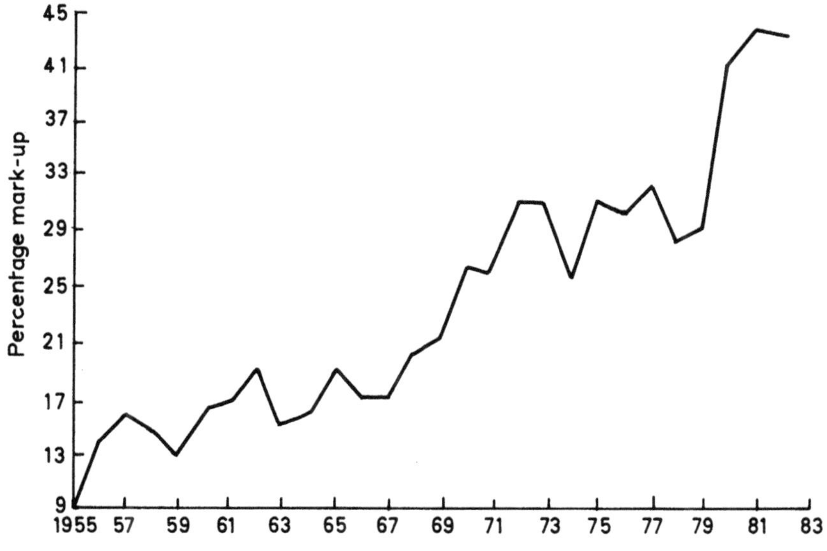

Fig. 5.12 Union mark-up (union/non-union wage differential)

relationship to shift after 1980, both because of the passing of the new trade union laws, and because union membership dropped (dramatically) for the first time since the war; in particular, the lags in effect could well lengthen as unions evaluate and adjust to the new environment. Secondly, they used a revised series for union penetration which suggest a lag of eight quarters in effect in the original sample; with this lag, the extended sample result is no different (Figure 5.11 shows our original measurement and the Henry *et al.* revision). Nevertheless, it is quite likely that the lag will lengthen further when the sample is extended more. Thirdly, there is fortunately another series available to proxy union power, the estimate of the union mark-up produced from earnings surveys by Layard *et al.* (1978, updated data courtesy of the authors and shown in Fig. 5.12). When this is used, the (significant) relationship does not shift when the sample is extended, which is what you would expect, since it is a direct measure of what unions have actually done in wage bargaining. In sum it is obvious that unions have driven up real wages as their power increased up to 1980. Quite when the drastic recent reduction in their powers will work through to reduced union mark-ups is something our sample cannot tell us and a matter of judgement at this point.

While the critics have disputed particular estimates in the work, it has nevertheless sufficiently affected the thinking of Keynesians for similar mechanisms to be incorporated in disequilibrium models (Layard and Nickell, 1985). From a policy viewpoint this represents an important convergence of view on the determinants of long run unemployment.

5.5 CONCLUSIONS

This chapter has attempted to produce a theory of the 'natural rate' for the open economy. The interaction with the balance of payments on current account via a stock flow mechanism has been argued to reinforce the direct effects of labour market intervention on unemployment and output. This interaction parallels the feedback onto the labour market through the budget constraint (output = spending) in a closed economy subject to the same intervention. Our major findings were that there is a significant and powerful total elasticity of real benefits on unemployment (operating through higher real wages) substantially higher than other postwar estimates in so far as these are comparable. Tax rates on employers and employees have analogous impacts, though the elasticities are much lower. Finally, and perhaps most strikingly, we find that in the past two decades union monopoly power has increased significantly and caused a substantial rise in real wages, with corresponding unemployment. The total elasticity of unemployment to the unionization rate (our index of union monopoly power) is no less than $5\frac{1}{2}$.

The natural rate of unemployment in the UK in 1980 is estimated to be of the order of $7\frac{1}{4}\%$ or around $1\frac{3}{4}$ million. The analysis of this chapter suggests that it can be lowered substantially by measures (such as those discussed in Minford *et al.*, 1983) to reduce real benefits, labour tax rates and union monopoly power.

ACKNOWLEDGEMENT

I am grateful to the editors of *Oxford Economic Papers* for permission to use material from Minford (1983). The many debts I reported there apply equally to this chapter.

REFERENCES

Bain, G. and Price, R. (1980) *Profiles of Union Growth: A Comparative Statistical Portrait of 8 Countries*, Blackwell, Oxford.

Batchelor, R. A. and Sheriff, T. D. (1980) Unemployment and unanticipated inflation in post war Britain, *Economica*, **47**, 179–92.

Benjamin, D. and Kochin, L. (1979) Searching for an explanation of unemployment in inter war Britain, *Journal of Political Economy*, **87**, No. 3, 441–70.

Benjamin, D. and Kochin, L. and Critics (1982) Unemployment and unemployment benefits in twentieth century Britain: a reply to our critics, and preceding critics' comments, *Journal of Political Economy*, **90**, 369–436.

Brookes, S. and Henry, S. G. (1983) Reestimation of the National Institute model, *National Institute Economic Review*, **103**, 62–70.

Godfrey, L. G. (1978) Testing for higher order serial correlations in regression equations when the regressors include lagged dependent variables, *Econometrica*, **46**, 1303–10.

Henry, S. G. B., Payne, J. M. and Trinder, C. (1985) Real wages and unemployment: the role of unemployment, Social Security benefits and unionization, *Oxford Economic Papers*, **37**, 330–38.

Hines, A. G. (1964) Trade unions and wage inflation in the United Kingdom, 1893–1961, *Review of Economic Studies*, **31**, 221–52.

Holden, K. and Peel, D. A. (1979) The determinants of the unemployment rate: some empirical evidence, *The Statistician*, **20**, No. 2, 101–7.

Holden, K. and Peel, D. A. (1981) Unemployment and the replacement ratio – some reduced form estimates for the UK, *Economics Letters*, **8**, 349–54.

Junankar, P. N. (1981) An econometric analysis of unemployment in Great Britain, 1952–75, *Oxford Economic Papers*, **33**, 387–400.

Lancaster, T. (1979) Econometric methods for the duration of unemployment, *Econometrica*, **47**, 939–56.

Layard, R., Metcalf, D, and Nickell, S. (1978) The effect of collective bargaining on relative and absolute wages, *British Journal of Industrial Relations*, **16**, November, 287–302.

Layard, R., and Nickell, S. J. (1985) The causes of British unemployment, *National Institute Economic Review*, No. 111 (February) p. 62.

Lewis, H. G. (1963) *Unionism and Relative Wages in the United States*, University of Chicago Press.

Lucas, R. E. and Rapping, L. S. (1969) Real wages, employment and inflation, *Journal of Political Economy*, **77**, 721–54.

MacKay, D. E. and Reid, G. L. (1972) Redundancy unemployment and manpower policy, *Economic Journal*, **82**, 1256–72.

Maki, D. and Spindler, A. (1975) The effect of unemployment compensation on the rate of unemployment in Great Britain, *Oxford Economic Papers*, **27**, 440–54.

Minford, A. P. L. (1980) A rational expectations model of the UK under fixed and floating exchange rates, in *The State of Macroeconomics*, Carnegie Rochester Conference Series on Public Policy **12**, 293–355.

Minford, A. P. L. (1983), Labour Market Equilibrium in an open economy, *Oxford Economic Papers*, **35**, November supplement, 207–44.
Minford, A. P. L. (1984) Response to Nickell, *Economic Journal*, **94**, 954–9.
Minford, A. P. L. (1985) Response to Henry, Payne and Trinder, *Oxford Economic Papers*, **37**, No. 2, 339–43.
Minford, A. P. L., Davies, D. H., Peel, M. J. and Sprague, A. (1983) *Unemployment – Cause and Cure*, Ist edn, Martin Robertson, Oxford.
Minford, A. P. L. and Hilliard, G. (1978) The cost of variable inflation, in *Contemporary Economic Analysis* (eds M. J. Artis and A. R. Nobay), Croom Helm, London.
Mulvey, C. (1976) Collective agreements and related earnings in UK manufacturing in 1973, *Economica*, **43**, No. 172, 419–27.
Muth, J. F. (1961) Rational expectations and the theory of price movements, *Econometrica*, **29**, 315–35.
Nickell, S. (1979a) The effects of unemployment and related benefits on the duration of unemployment, *Economic Journal*, **89**, March, 34–49.
Nickell, S (1979b) Estimating the Probability of Leaving Unemployment, *Econometrica*, **47**, 1249–66.
Nickell, S. (1981) The demand for labour function in UK manufacturing, paper presented to Treasury Academic Panel, Centre for Labour Economics, London School of Economics.
Nickell, S. J. (1984) A review of unemployment: cause and cure by Patrick Minford, with David Davies, Michael Peel and Alison Sprague, *Economic Journal*, **94**, 946–53.
Oswald. A. J. (1982) The microeconomic theory of the trade union, *Economic Journal*, **92**, 576–95.
Parsley, C. J. (1980) Labour union effects on wage gains: a survey of recent literature, *Journal of Economic Literature*, **18**, 1–31.
Pencavel, J. H. (1974) Relative wages and trade unions in the United Kingdom, *Economica*, **41**, 194–210.
Purdy, D. L. and Zis, G. (1973) Trade unions and wage inflation in the UK: a reappraisal, in *Essays in Modern Economics* (ed. M. Parkin), Longman, 1973, 294–327; reprinted in *Inflation and Labour Markets* (eds D. Laidler and D. Purdy), Manchester University Press, 1974.
Sargan, J. D. (1964) Wages and prices in the United Kingdom, in *Econometric Analysis for National Economic Planning* (eds P. E. Hart, G. Mills and J. K. Whittaker) Butterworth, London.
Sargent, T. J. (1976) A classical macroeconomic model of the United States, *Journal of Political Economy*, **84**, 207–38.
Sargent, Thomas J. (1979) *Macroeconomic Theory*, Academic Press, New York.
Smith, R. (1983) On the classical nature of the Wu-Hausman statistics for the independence of stochastic regressors and disturbance, *Economics Letters*, **11**, No 4, 357–64.
Treble, J. G. (1982) Does the union/non-union wage differential exist? Paper given to 1982 Association of University Teachers of Economics, Conference at Surrey University (mimeo), University of Hull.
Wallis, K. F. (1980) Econometric implications of the rational expectations hypothesis, *Econometrica*, **48**, 49–72.

6

Imperfect competition and the labour market

STEPHEN NICKELL

6.1 INTRODUCTION

It has long been appreciated that wage determination is crucial to the behaviour of the macroeconomy. More recently, however, it has been recognized that the aggregate relationship between real wages and employment depends not only on the workings of the labour market but also on the pricing behaviour of firms. In this chapter, we are concerned to illustrate the importance of conditions in both the product market and the labour market in the determination of macroeconomic outcomes. We develop a simple, one good, model of the economy looking first at the behaviour of firms and then at the workings of the labour market. We then consider the macroeconomy as a whole and conclude by looking at some empirical results.

6.2 PRICE AND EMPLOYMENT BEHAVIOUR

We begin by considering a single industry, labelled *i*. A firm in the industry operates in each period with a predetermined capital stock, K_{ij}, where *j* is the firm index. Any investment which the firm undertakes during the period is simply added to the capital stock for next period and has no implications for the current level of productive capacity. We suppose that firms set prices, output and employment at the beginning of the period on the basis of expected demand. Wages are also determined at this time. If demand turns out to be different from that expected, then the difference between demand and output is taken up by a change in inventory holdings. Each firm in the industry produces value-added output using capital and labour via a constant-returns production function of the form

$$Y_{ij} = F(A_i N_{ij}/K_{ij})K_{ij}, \qquad F' > 0, \qquad F'' < 0, \tag{6.1}$$

where N_{ij} = employment, Y_{ij} = value-added output, and A_i = labour-augmenting technical progress. It is convenient to have the inverse form of this production relation available, namely

$$A_iN_{ij}/K_{ij} = G(Y_{ij}/K_{ij}), \qquad G = F^{-1}, \qquad G' > 0, \qquad G'' > 0. \tag{6.2}$$

Firms in the industry behave in the following way. First, the value-added output price is set on the basis of expected demand. Then, given the price, output is produced to satisfy expected demand and employment is fixed via the production function.

In order to describe pricing behaviour, we must specify the demand facing the industry. Let us first denote by $\bar{Y}$ the level of output in the economy corresponding to full utilization of resources. Then we may define real aggregate demand as $e^{\sigma}\bar{Y}$, so that σ represents a measure of aggregate demand relative to full utilization output. So if P_i is the price of value-added output from industry i and P is the aggregate equivalent, then we may specify the expected demand for industry i, Y_i^{ed}, as

$$Y_i^{ed} = \omega_i(P_i/P^e)^{-\theta_i}e^{\sigma^e}\bar{Y}, \qquad \theta > 1 \tag{6.3}$$

where ω_i is the share of total aggregate demand falling to the ith industry when $P_i = P^e$.

Price behaviour we suppose to have the following characteristics. First, there is a uniform price across the industry, with demand being allocated to each firm in proportion to its initial capital stock. In the light of this, the output–capital ratio is the same for all firms in the industry and hence Equation 6.2 implies that the labour–capital ratio is also uniform. So there exists an industry wide production relation corresponding to Equation 6.2, namely

$$A_iN_i/K_i = G(Y_i/K_i) \tag{6.4}$$

where

$$N_i = \sum_j N_{ij}, \quad K_i = \sum_j K_{ij} \quad \text{and} \quad Y_i = \sum_j Y_{ij}.$$

Second, the price is related to the joint monopoly price, P_i^*. In general, this relationship will not be fixed over the cycle but will vary in ways we shall discuss later. In order to define this joint monopoly price, we may note that marginal cost is given by

$$MC = \frac{W_i}{A_i} G'(Y_i/K_i) \tag{6.5}$$

where W_i represents labour cost per period (that is, the gross wage plus employment taxes). The joint monopoly price is then given by

$$P_i^* = \frac{\theta_i}{\theta_i - 1}\frac{W_i}{A_i} G'(Y_i/K_i) \tag{6.6}$$

since θ_i is the elasticity of demand. The question now arises as to how the actual output price relates to P_i^*. The Cournot–Nash solution yields a price which is fixed proportion of the monopoly price, since it generates a fixed mark-up on marginal cost which depends on the Herfindahl index of concentration in the industry.[1] However, other theories indicate that deviation of industry prices from the joint monopoly level are sensitive to the cycle. In oligopolistic industries, Rotemberg and Saloner (1986) provide a theoretical basis and empirical evidence for the notion that oligopolies find implicit collusion more difficult when demand is high. The idea here is simply that the benefits of deviating from collusive behaviour are more likely to outweigh the costs as demand rises, the analysis being undertaken in the context of a repeated game. This would imply that the mark-up on marginal cost moves countercyclically.[2] In industries with relatively free entry and exit, on the other hand, only average cost pricing can limit the threat of entry (see Mirman *et al.* (1986) for example). Given increasing marginal costs, this would again tend to lead to a countercyclical mark-up on marginal cost. The implication of this discussion is that we might expect actual prices, P_i, to satisfy

$$P_i = f_i(\sigma^e)P_i^* = f_i(\sigma^e)\frac{\theta_i}{\theta_i - 1}\frac{W_i}{A_i}G'(Y_i/K_i) \tag{6.7}$$

where $0 < f_i \leqslant 1$ and $f_i' \leqslant 0$.

Once P_i has been determined, then industry output is set in order to satisfy expected demand. So, from Equation 6.3 we have

$$Y_i = \omega_i(P_i/P^e)^{-\theta_i}e^{\sigma^e}\bar{Y} \tag{6.8}$$

and industry employment is determined by the production relation 6.4. Thus given W, P^e, K_i, σ^e, this industry model determines P_i, Y_i and N_i.

One particular point worth noting is the relationship between price determination and the level of employment. To see this, note that since $G = F^{-1}$, we have

$$G'(Y_i/K_i) = [F'(A_iN_i/K_i)]^{-1} \tag{6.9}$$

which is, in fact, another way of writing the technology. Using this, we may rewrite the price equation 6.7 as

$$F'(A_iN_i/K_i) = f_i(\sigma^e)\frac{\theta_i}{\theta_i - 1}\frac{W_i}{A_iP_i} \tag{6.10}$$

This is the marginal revenue product condition and reveals that the marginal product of labour is equal to the product wage times the mark-up of prices on marginal cost. So, for example, if we have a Cournot–Nash industry with a constant elasticity demand function, $f_i\theta_i/(\theta_i - 1)$ is constant and the relationship between employment and the product wage in that industry is exactly the same

[1] In fact the formula is $P_i = (\theta_i/\theta_i - H)MC$ where H is the Herfindahl index of concentration $\sum_j(Y_{ij}^2/Y_i^2)$. See Cowling and Waterson (1976), for example.

[2] Bils (1985) argues that the elasticity of demand is higher in booms as more 'unattached' customers search for new products. This has precisely the same implication.

as for a competitive industry, aside from the constant. Indeed, in a macroeconomic context, for all practical purposes an economy consisting of Cournot–Nash industries with constant elasticity demand will behave in the same way as an otherwise similar economy with a competitive product market. Imperfect competition, *per se*, makes no odds. However, if imperfect competition implies that the price mark-up on marginal cost varies countercyclically, then we can see immediately from Equation 6.10 that this will cause the standard relationship between the marginal product of labour and the product wage to shift outwards in booms and inwards in slumps. Not surprisingly, this has profound implications for macroeconomic behaviour, in particular the response of real wages and employment to various kinds of shock.

Returning to the development of our model, we next want to add-up over the industries to obtain an aggregate model of the firm sector. In keeping with the 'one good' nature of the standard macro model, we simply suppose that all industries produce the same good, have the same technological characteristics and face uniform wages. So we have $f_i = f$, $\theta_i = \theta$, $A_i = A$, $P_i = P$, $W_i = W$. As a consequence the output–capital and labour–capital ratios are uniform (see Equations 6.7, 6.4) and we may aggregate Equations 6.4, 6.7 and 6.8 (noting that $\sum \omega_i = 1$) to obtain

$$AN/K = G(Y/K) \qquad \text{production relation} \tag{6.11}$$

$$P/W = \frac{f(\sigma^e)\theta}{A(\theta - 1)} G'(Y/K) \qquad \text{pricing} \tag{6.12}$$

$$Y = (P/P^e)^{-\theta} e^{\sigma^e} \bar{Y} \qquad \text{demand} \tag{6.13}$$

where N, K, Y are economy-wide aggregates.

In order to tie up with our discussion of empirical work it is convenient to rewrite these equations in log-linear form as

$$n + a - k = a_0 + a_1(y - k) \tag{6.11'}$$

$$p - w = b_0 - b_1\sigma^e + b_2(y - k) - a \tag{6.12'}$$

$$y = -\theta(p - p^e) + \sigma^e + \bar{y} \tag{6.13'}$$

where lower case letters refer to logs and all parameters are positive. To give some idea of the size and meaning of these parameters it is easy to show that

$$a_1 = 1/S_L > 1, \quad b_2 = \frac{1 - S_L}{S_L \sigma_{KN}} \tag{6.14}$$

where $S_L = (AN/K)(F'/F)$, the 'share of labour' and σ_{KN} is the elasticity of substitution between labour and capital. The parameter b_1 simply captures the effect of the level of activity on the price mark-up on marginal cost. If we eliminate y between Equations 6.12′ and 6.13′ we have the standard pricing equation

$$p - w = b_0 + (b_2 - b_1)\sigma^e - b_2\theta(p - p^e) + b_2(\bar{y} - k) - a \tag{6.15}$$

with the price mark-up on wages depending on demand, σ^e, price surprises

$(p - p^e)$ and trend productivity $(b_2(\bar{y} - k) - a)$. The demand effect consists of two offsetting forces, the positive one, b_2, arising from increasing marginal costs and the negative one $(-b_1)$ arising from the possible fall in the mark-up on marginal cost as demand increases. If these exactly cancel out, then we have the standard 'normal cost' pricing model. The negative impact of price surprises essentially reflects nominal inertia in price setting. It arises in this model because firms may not adjust their expectations about competitors' prices when wage costs increase. This leads to a squeeze of mark-up of prices on wages when nominal magnitudes run ahead of expectations, typically when inflation is increasing. Any element of 'historic cost' pricing will, of course, exacerbate this tendency. Finally, the trend productivity effect is more easily understood if we can eliminate the term in full utilization output, $\bar{y}$, which is, in any event, rather difficult to observe. Given the production relation (Equation 6.11′) it seems natural to define full utilization output by

$$l + a - k = a_0 + a_1(\bar{y} - k) \tag{6.16}$$

where $l = \log L$, L = labour force. If we substitute this into our price equation, Equation 6.15, we obtain

$$p - w = \left[b_0 - \frac{b_2 a_0}{a_1} \right] + [b_2 - b_1]\sigma^e - b_2\theta[p - p^e] - \frac{b_2}{a_1}(k - l) - \left[1 - \frac{b_2}{a_1} \right] a \tag{6.17}$$

So trend productivity effects on prices arise from two forces, capital accumulation reflected in the $(k - l)$ term and labour augmenting technical progress, a. Equation 6.17 forms the basis of our empirical price equation discussed later.

For empirical analysis it is convenient to use an employment equation based on the marginal productivity relation, Equation 6.10, which we have already discussed. This may be obtained by eliminating y from Equations 6.11′ and 6.12′ and rearranging to yield

$$n - k = \left[a_0 - \frac{a_1 b_0}{b_2} \right] + \frac{a_1 b_1}{b_2}\sigma^e - \frac{a_1}{b_1}(w - p) + \left[\frac{a_1}{b_1} - 1 \right] a \tag{6.18}$$

This serves as the foundation for our empirical employment functions. One final point is worth emphasizing. If the price mark-up on marginal cost is constant, then $b_1 = 0$ and Equation 6.18 has the form of a competitive labour demand function. But in this case, prices must be increasing in demand (unless, of course, marginal costs are constant $(b_2 = 0)$ which is hardly realistic when the capital stock is fixed). Conversely if prices are unaffected by demand b_1 must be positive (since $b_1 = b_2$) and the employment relation cannot have the competitive form.

6.3 WAGE DETERMINATION

The next step in constructing our model is to consider the determination of wages. It is highly probable that in any economy the system of wage determination differs across sectors. In some sectors wages may be competitively determined, in others, firms set wages and in yet others they arise out of bargaining between employers and unions. There are a large number of theories of wage determination covering all these mechanisms and many of them are described in Stiglitz (1986) and Johnson and Layard (1986). As far as the structure of the wage equation is concerned, it is fortunate that these theories have broadly similar implications. In this section we present a derivation of the wage equation which captures many of the salient features of these theories without being too specific as to which one is dominant.

We begin with the marginal revenue product condition (Equation 6.18) which may be rewritten at the industry level as

$$n_i - k_i = \alpha_0 + \alpha_1 \sigma^e - \alpha_2(w_i - p_i) + \alpha_3 a_i \tag{6.19}$$

where $\alpha_0 = a_0 - a_1 b_0/b_2$, $\alpha_1 = a_1 b_1/b_2$, $\alpha_2 = a_1/b_2$, $\alpha_3 = (a_1/b_2) - 1$. Wages in industry i are determined at the beginning of the period. Now define w_i^I as that wage which would have to rule if $\bar{n}_i$ individuals were to be employed in the industry in the long run at some average level of demand $\bar{\sigma}$. Thus w_i^I is given by

$$w_i^I = p_i^e + \frac{1}{\alpha_2}[(k_i - \bar{n}_i) + \alpha_0 + \alpha_1\bar{\sigma} + \alpha_3 a_i] \tag{6.20}$$

w_i^I may be thought of as the 'insider' wage, that is the wage which would stabilize long-run employment in the industry at $\bar{n}_i$. A pure 'insider' theory of wage setting along the lines of Blanchard and Summers (1986), for example, would set $\bar{n}_i = n_{i-1}$, arguing that this is the sole force behind wage determination in a world dominated by unions whose prime concern is for their existing members. A somewhat more expansive union might also attach some weight to recently unemployed workers from industry i and we would then have

$$\bar{N}_i = N_{i-1} + \omega U_{i-1} \tag{6.21}$$

where $0 \leqslant \omega \leqslant 1$ and U_i refers to those unemployed who are 'attached' to industry i. Equation 6.21 then implies

$$\begin{aligned} \bar{N}_i &= L_{i-1} + (\omega - 1) U_{i-1} \\ &= L_{i-1}[1 + (\omega - 1) u_{i-1}] \end{aligned}$$

where L_i is the labour force associated with industry i and u_i is the unemployment rate among this group. This enables us to write

$$\bar{n}_i = l_{i-1} + (\omega - 1) u_{i-1} \tag{6.22}$$

which is an obvious generalization of the previous story. It can also be argued that managers are not averse to wage setting along these lines since it leads to a quiet life and reflects the somewhat cosy relationship between managers and unions that is seen as a feature of many sectors of industry, particularly in Britain.

It seems likely, however, that even the most cosy of industries cannot be immune from outside forces and we may define the 'outsider' wage as reflecting this fact.

This wage, w_i^{O}, reflects the outside opportunities of the workforce or the wage that must be paid to attract outside workers. This depends on wages that are expected to rule elsewhere modified by the chances of obtaining outside employment, the relative attractiveness of the unemployed state and the extent to which the unemployed are both suitable and willing to work in the industry. Thus we may define w_i^{O} as

$$w_i^{\mathrm{O}} = \bar{w}_i^{\mathrm{e}} - g(u) + z^1, \qquad g' > 0 \tag{6.23}$$

where $\bar{w}_i^{\mathrm{e}}$ is the expected outside wage, u is the general unemployment rate and z^1 is a set of variables which reflects both the extent to which the unemployed are unsuitable for work in the industry (for example, mismatch variables) and the attractiveness of the unemployed state (for example, the replacement ratio).

The actual wage outcome will reflect both 'insider' and 'outsider' forces. Two other factors might also be important. First, an industry with a more powerful union may simply have a higher wage level, *ceteris paribus*. Second, workers may resist wage adjustment associated with changes in the wedge between product wages and consumption wages (post-tax wages deflated by the RPI). In other words, if tax changes, for example, raise the product-wage relative to the consumption wage, workers may resist the reduction in the consumption wage which is necessary to stabilize employment. Elements of the wedge are employers' labour taxes, income taxes, excise taxes and the relative price of imports, the latter arising because the product price refers to value-added in our model whereas the RPI refers to a large segment of gross output. This suggests an industry wage equation of the form

$$w_i = \lambda_0 + \lambda_1 w_i^{\mathrm{I}} + (1 - \lambda_1) w_i^{\mathrm{O}} + z^2$$

where z^2 contains union power and elements of the wedge. Substituting for w^{O}, w^{I} using Equation 6.22 yields

$$w_i = \lambda_0 + \lambda_1 p_i^{\mathrm{e}} + (1 - \lambda_1)\bar{w}_i^{\mathrm{e}} + \frac{\lambda_1}{\alpha_2}\{[k_i - l_{i-1}] + (1 - \omega)u_{i-1}$$
$$+ \alpha_0 + \alpha_1 \bar{\sigma} + \alpha_3 a_i\} - (1 - \lambda_1) g(u) + (1 - \lambda_1) z^1 + z_2$$

In order to aggregate to the whole economy level we make the same assumptions as in the previous section and obtain

$$w = \lambda_0 + \lambda_1 p^{\mathrm{e}} + (1 - \lambda_1) w^{\mathrm{e}} + \frac{\lambda_1}{\alpha_2}\{[k - l_{-1}] + (1 - \omega)u_{-1}$$
$$+ \alpha_0 + \alpha_1 \bar{\sigma} + \alpha_3 a\} - (1 - \lambda_1) g(u) + (1 - \lambda_1) z^1 + z^2$$

This may be rewritten as

$$w - p = \gamma_0 - [(p - p^{\mathrm{e}}) + \gamma_1 (w - w^{\mathrm{e}})] + \frac{1}{\alpha_2}(k - l) + \frac{\alpha_3}{\alpha_2} a$$
$$- f(u, u_{-1}, \Delta l) + \bar{z} \tag{6.24}$$

where

$$\gamma_0 = \lambda_0/\lambda_1 + (\alpha_0 + \alpha_1\bar{\sigma})/\alpha_2, \quad \gamma_1 = (1-\lambda_1)/\lambda_1,$$

$$f(u, u_{-1}, \Delta l) = \gamma_1 g(u) - \frac{(1-\omega)}{\alpha} u_{-1} - \frac{1}{\alpha_2}\Delta l, \quad \bar{z} = \frac{1}{\lambda_1}[z^2 + (1-\lambda_1)z^1).$$

Several points are worth noting about this wage equation. First, it has a relatively general form and is consistent with more or less all models of wage setting behaviour. Second, it contains four standard groups of variables. Nominal inertia captured by price and wage surprises, trend productivity effects which are related to those on employment and price determination via the α_2, α_3 parameters, conditions in the labour market, where the positive lagged unemployment effect is of particular interest, and general wage pressure arising from the $\bar{z}$ variables.

6.4 THE LABOUR MARKET IN A MACROMODEL

Having set out the price and employment behaviour of firms and a model of wage determination, it now remains to complete a macromodel and discuss the role of the labour market in determining its response to shocks. We consider a model conditional on the level of aggregate resources (capital and labour force) and on technology (technical progress). Thus we omit the interactions between macropolicy, levels of demand and the rates of capital accumulation and technical progress. The determinants of long run growth rates are, of course, matters of great interest but here we wish to concentrate on the rate of utilization of available resources rather than their absolute levels.

To construct our macromodel, we first gather together the supply-side equations which we have already discussed in some detail. Before writing them down, however, we must be more explicit about one point which is of some importance in the context of an open economy. One of the wage pressure variables, $\bar{z}$ in Equation 6.24, is the real price of imports which is part of the wedge between the consumption and product wages. If the real price of imports rises, then the consumption wage must fall if the product wage is to remain unchanged. Real wage resistance will then imply upward pressure on the product wage. We may write the real price of imports as $e + p_m - p$, where p_m is the (log) import price in overseas currency and e is the exchange rate (measured so that a rise in e represents a depreciation of the home currency). Then we have

$$\begin{aligned} e + p_m - p &= (p_m - p^*) + (e + p^* - p) \\ &= (p_m - p^*) + c, \text{ say,} \end{aligned}$$

where p^* is the world price of output in overseas currency and c is competitiveness. Thus the real price of imports can be divided into the exogenous world terms of trade between home importables and world output, $(p_m - p^*)$, and endogenous output price competitiveness, c. In the light of this, it is worthwhile replacing $\bar{z}$ in the wage equation by c and a set of exogenous wage pressure variables z. So our set of supply-side equations are

Demand (6.13′)

$$y = -\theta(p - p^e) + \sigma^e + \bar{y} \tag{6.25}$$

Employment (6.18, 6.19)

$$n - k = \alpha_0 + \alpha_1\sigma^e - \alpha_2(w - p) + \alpha_3 a \tag{6.26}$$

Prices (6.17)

$$p - w = \beta_0 + \beta_1\sigma^e - \beta_2(p - p^e) - \frac{1}{\alpha_2}(k - l) - \frac{\alpha_3}{\alpha_2}a \tag{6.27}$$

Wages (6.24)

$$w - p = \gamma_0 - \gamma_2(p - p^e) + \frac{1}{\alpha_2}(k - l) + \frac{\alpha_3}{\alpha_2}a$$

$$- f(u, u_{-1}, \Delta l) + \gamma_3 c + z \tag{6.28}$$

Identity

$$u = l - n \tag{6.29}$$

Note that we have eliminated $(w - w^e)$ from the wage equation. We are able to do this since the price equation implies a linear relationship between $(p - p^e)$ and $(w - w^e)$.[3]

The demand side we shall use assumes both floating exchange rates and perfect capital mobility/substitutability.[4] So we have

IS curve $\quad y^d = \sigma(x, r - \dot{p}^e, y^*, c) + \bar{y}$ (6.30)

LM curve $\quad m - p = L(r, y^d)$ (6.31)

Arbitrage condition $r = r^* + \dot{e}^e$ (6.32)

Identity $\quad c = e + p^* - p$ (6.33)

y^d = aggregate real demand, x = fiscal policy variables, r = domestic nominal interest rate, y^* = world activity, m = money stock, p^* = world output price, r^* = world nominal interest rate, e = nominal exchange rate, $\dot{p}^e$ = expected inflation rate, $\dot{e}^e$ = expected depreciation (remember a rise in e implies exchange rate depreciation from the domestic viewpoint). In order to use this framework, we must also specify how exchange rate expectations are determined. We suppose that

$$\dot{e}^e = \dot{p}^e - \dot{p}^{*e} - \delta(c - c^*), \quad 1 > \delta > 0$$

where c^* is the long-run 'equilibrium' level of competitiveness which, for the moment, we assume to be constant. Thus the expected rate of depreciation depends on the expected inflation differential minus a factor representing the extent to which competitiveness is above its long-run level.

[3] Implicit in our model is that technology conditions are known in advance. So if we take expectations of the price function and subtract the resulting equation from the original, we obtain $(p - p^e) - (w - w^e) = -\beta_2(p - p^e)$.

[4] These demand side assumptions are chosen partly for realism and partly for simplicity. Any combination of fixed/floating and perfect/imperfect capital mobility could equally well be used.

If we now combine this with Equation 6.32, we can use it to eliminate r from the IS, LM curves to obtain

$$\text{IS:}\qquad y^{d}=\sigma(x, r^{*}-\dot{p}^{*e}-\delta(c-c^{*}), y^{*}, c)+\bar{y} \tag{6.34}$$

$$\text{LM:}\quad m-p=L(\dot{p}^{e}+(r^{*}-\dot{p}^{*e})-\delta(c-c^{*}), y^{d}) \tag{6.35}$$

How does the model then operate? In order to keep things simple, suppose that the level of real demand expected by firms is the actual level which turns up.[5] That is

$$\sigma^{e}=\sigma(x, r^{*}-\dot{p}^{*e}-\delta(c-c^{*}), y^{*}, c) \tag{6.36}$$

Furthermore note that all the analysis is conditional on resources and technology $(k, l, \bar{y}, a)$. Then, we can first look at the short run when p^{e} is predetermined. Starting from the IS curve (Equation 6.34), this determines aggregate demand as a function of exogenous factors and competitiveness. Since the exogenous factors include fiscal policy variables, aggregate demand can be set at any predetermined level whatever the level of competitiveness. So, in the short run, we can view demand as exogenous. Given demand, the supply side (6.25, ..., 6.29) will generate output, employment, unemployment, the real wage and prices for any given level of competitiveness and exogenous wage pressure. The LM curve (Equation 6.35) will pin down a further relationship between prices and competitiveness at given demand, thereby determining all the endogenous variables in the model $(y, n, u, w-p, p, c)$. Furthermore, note that given demand, the LM relationship between c and p can be shifted at will by manipulating monetary policy (changing m). This is useful for expositional purposes because it reveals that monetary policy can be adjusted to generate any desired level of competitiveness. So for short run analysis we can effectively treat competitiveness as exogenous as well as demand.

What of the long run? This we define as the state where expectations are fulfilled, that is $p=p^{e}$. It is clear from the supply side of the model that for given competitiveness and wage pressure, there is only one level of demand, σ, which will generate $p-p^{e}=0$. Indeed if we set $p-p^{e}=0$ in Equations 6.27 and 6.28 then these two equations plus Equations 6.26 and 6.29 will yield that level of σ along with the corresponding values of employment, n, the real wage, $w-p$, and unemployment, u. All this, however, is conditional on competitiveness, c, which, as we have already remarked, can be manipulated by monetary policy for a given level of demand. This appears to imply that the natural rate of unemployment can, itself, be manipulated by government policy. This is not, however, correct. In the long run, competitiveness cannot be set at will because the balance of trade will exert a constraint. If, for the sake of argument, we impose a condition of trade balance in the long run,[6] this will fix the relationship between demand and competitiveness

[5] This merely simplifies the discussion. It has no profound implications.

[6] Trade balance does not, of course, have to be zero in the long run. We could tolerate some fixed level of accumulation of foreign assets either by the domestic or foreign country if there were differential rates of capital accumulation, for example. This would have the same implications so we stick to the simplest assumption.

thereby pinning down all the real variables in the economy. Writing this formally, suppose the trade balance condition is

$$c = \omega_0 + \omega_1 \sigma - z_c, \quad \omega_1 > 0 \tag{6.37}$$

where z_c are a set of exogenous variables which would include world activity, y^*, and North Sea oil, both of which would tend to reduce competitiveness. Then the supply side plus (Equation 6.37) will determine a unique natural rate for any given level of wage pressure (and z_c).[7] The demand side simply enables us to read off the level of fiscal and monetary policy instruments (x, m) consistent with long-run equilibrium.

In the light of this discussion, we can consider the workings of the model simply by looking at the supply-side equations. If we write the trade balance level of competitiveness as c^*, then we can use Equation 6.37 to rewrite the wage equation, Equation 6.28, as

$$w - p = \gamma_0 - \gamma_2(p - p^e) + \frac{1}{\alpha_2}(k - l) + \frac{\alpha_3}{\alpha_2} a - f(u, u_{-1}, \Delta l)$$

$$+ \gamma_3(c - c^*) + \gamma_3\omega_0 + \gamma_3\omega_1\sigma - \gamma_3 z_c + z \tag{6.38}$$

where we can treat $c - c^*$ as exogenous in the short run but as zero in the long run. The two key equations are now this and the price equation (Equation 6.27), the only problem being that they contain two different measures of the level of activity, σ and u. Since we want to look at things in the context of the labour market, we can use the marginal revenue product condition (Equation 6.26), which relates product market activity to labour market activity, to eliminate σ from Equations 6.38 and 6.27 so as to view everything from the side of the labour market. This yields

$$\text{Prices:} \quad p - w = \beta_{10} - \beta_{11}(p - p^e) - \beta_{12}u - \text{prod} \tag{6.39}$$

$$\text{Wages:} \quad w - p = \gamma_{10} - \gamma_{11}(p - p^e) - g(u, u_{-1}, \Delta l) + \text{prod} + \tilde{z} \tag{6.40}$$

$$\left(\begin{array}{l} \beta_{10} = (\alpha_1\beta_0 + \beta_1\alpha_0)/\Delta_1,\ \beta_{11} = \alpha_1\beta_2/\Delta_1,\ \beta_{12} = \beta_1/\Delta_1,\ \Delta_1 = \alpha_1 + \beta_1\alpha_2 \\ \gamma_{10} = (\gamma_0\alpha_1 + \gamma_3\omega_0\alpha_1 - \gamma_3\omega_1\alpha_0)/\Delta_2,\ \gamma_{11} = \alpha_1\gamma_2/\Delta_2,\ \Delta_2 = \alpha_1 - \gamma_3\omega_1\alpha_2 \end{array}\right)$$

where

$$\text{prod} = \frac{1}{\alpha_2}(k - l) + \frac{\alpha_3}{\alpha_2} a, \text{ the trend productivity term,}$$

$$\tilde{z} = [z - \gamma_3 z_c + \gamma_3(c - c^*)]/\Delta_2, \text{ the wage pressure term,}$$

$$g(u, u_{-1}, \Delta l) = (f(u, u_{-1}, \Delta l) + \gamma_3\omega_1 u)/\Delta_2, \text{ the unemployment terms.}$$

[7] The model therefore has the 'natural rate' property despite the obvious absence of market clearing or competitive assumptions. It is, of course, possible to generate alternative models where the no-surprise equilibrium can be influenced by fiscal/monetary policy. This is straightforward in the case of two good models where government expenditure can be targeted differentially between the two goods (e.g. Hart, 1983 or Dixon, 1986a) or in one good models where the fiscal/monetary policy mix influences the firms' price mark-up (e.g. Wren–Lewis, 1985) or where the long-run real exchange rate can be manipulated at given demand (e.g. Carlin and Soskice, 1985). A model which is closely related to that presented here may be found in Dixon, 1986b.

This pared-down version of our original model is perfectly adequate for discussing the implications of demand and supply shocks. Furthermore, it has the great advantage of yielding the no-surprise or natural rate level of unemployment simply by setting $p = p^e$ and solving for u.

In order to see how the model operates, we consider, first, the consequences of an aggregate demand shock starting from a position of equilibrium ($p = p^e$). This is illustrated in Fig. 6.1, where we illustrate the price and wage equations in real-wage/unemployment space. An increase in real demand lowers unemployment, and the rise in activity in both the output and the labour market tends to raise prices relative to wages in the price equation and wages relative to prices in the wage equation. The only way in which these tendencies can be made consistent is via the positive price surprise brought about by a rise in inflation. This tends to offset the consequences of the rise in the level of activity on both sides of the market, leading us to a point such as B which is below the wage line and above the price line (note that the effect of a positive price surprise is to raise the price-determined real wage). So we have the standard result that a positive demand shock will lower employment and raise inflation but have an indeterminate impact on the real wage. However, we can be more precise on this latter point in certain special cases. If the product market is competitive, or more generally if the price mark-up on marginal cost is constant, the price line is essentially the competitive demand for labour schedule. B must lie on this line and the real wage must, therefore, fall. On the other hand, under strict normal cost pricing, $\beta_{12} = 0$ and the price line is horizontal. Under these circumstances the real wage must rise, since B will be above the horizontal price line.

Turning to the consequences of a supply shock, in Fig. 6.2 we illustrate the outcome of a rise in wage pressure, $\tilde{z}$. If real demand remains fixed, then we move to a point such as B, with a rise in unemployment and positive price surprises generated by rising inflation. The real wage has risen and we have the combination of inflation and 'classical' unemployment typical of such a shock. If the rise in wage

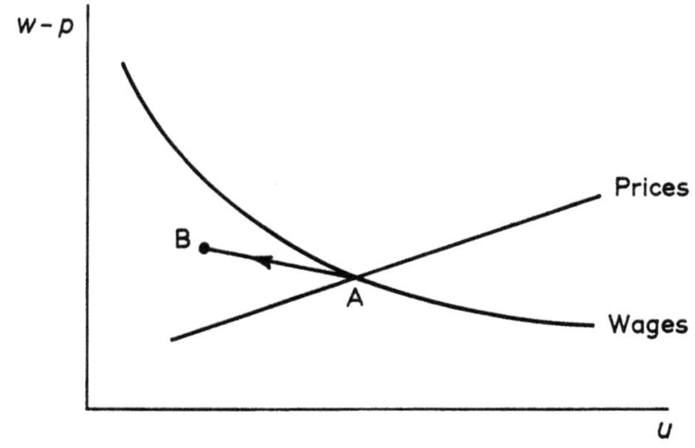

Fig. 6.1 The consequences of a positive demand shock

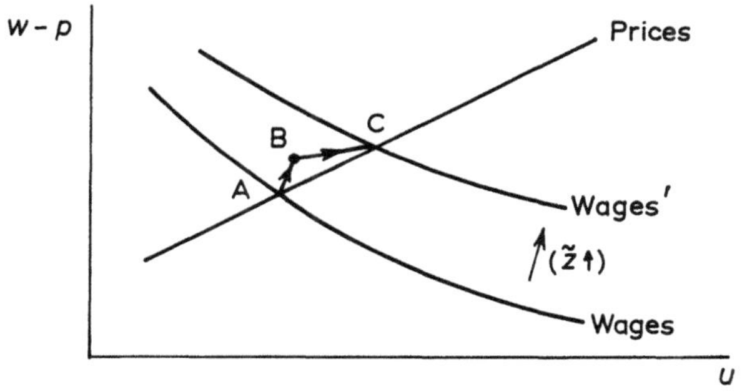

Fig. 6.2 The consequences of a rise in wage pressure

pressure is permanent, real demand must fall if inflation is to be stabilized. This may happen either autonomously (via real balance effects, for example) or as a result of a conscious policy shift. As a consequence, we move to a new equilibrium at C with higher 'equilibrium' unemployment and a real wage which will be higher to the extent that prices are influenced by demand. In the extreme case of normal cost pricing, the price line is horizontal and the real wage will revert back to its original level. The additional unemployment will then apparently be entirely 'Keynesian', although it has, in fact, been brought about by the rise in wage pressure.

It is clear from this analysis that the wage pressure variables are the key to the long-run analysis of unemployment. Furthermore, it is also clear that focusing on the real wage outcome will not be very useful in trying to understand what is happening. The real wage which finally emerges has as much to do with the pricing policy of firms as it has with labour market activity.

Another aspect of the workings of the economy which is highlighted by this minimodel is the unemployment-inflation tradeoff. If we eliminate the real wage from Equations 6.39 and 6.40 we obtain[8]

$$\beta_{12}u + g(u, u_{-1}, \Delta l) = (\beta_{10} + \gamma_{10}) - (\beta_{11} + \gamma_{11})(p - p^e) + \tilde{z}. \qquad (6.41)$$

This is the key equation since it reveals how unemployment is determined in the long run when price expectations are fulfilled ($p = p^e$) and, in the short run, how price surprises are associated with deviations from this long-run level. Let us now suppose that price surprises are associated with changes in the rate of inflation. When inflation is rising, then prices turn out to be higher than expected and vice versa. When inflation is completely stable, on the other hand, prices are accurately

[8] It is worth noting that our model has the property that trend productivity does not appear in Equation 6.41 because it appears with the same absolute coefficient in both the wage and price equation. This ensures that the level of productivity has no long run impact on unemployment. This is a sensible restriction in the sense that, were it not to hold, rising productivity through time would lead to persistently increasing or decreasing unemployment, which does not appear to have been the consequence of two centuries of economic growth.

forecast and $p - p^e = 0$. So assume that

$$p - p^e = \gamma\Delta^2 p. \qquad (6.42)$$

In reality, of course, we would expect there to be some dynamics in this equation but otherwise this appears to conform reasonably closely to the facts, at least in the recent past. So if we utilize Equation 6.42, then Equation 6.41 becomes

$$\beta_{12}u + g(u, u_{-1}, \Delta l) = (\beta_{10} + \gamma_{10}) - (\beta_{11} + \gamma_{11})\gamma\Delta^2 p + \tilde{z} \qquad (6.43)$$

When inflation is stable, unemployment is at the natural rate determined by the wage pressure variables, $\tilde{z}$ (with $c = c^*$). When unemployment deviates from the natural rate, this deviation is associated with a particular rate of increase or decrease of the rate of inflation. Furthermore, the form of this association depends crucially on the function $g(u, u_{-1}, \Delta l)$ which measures the impact of labour market activity on wage determination.

In order to see more clearly what is happening, we present in Fig. 6.3 the sort of picture which would emerge if $g(u, u_{-1}, \Delta l)$ was a simple linear function of current unemployment, i.e.

$$g(u, u_{-1}, \Delta l) = \gamma_{12}u \qquad (6.44)$$

The long-run natural rate of unemployment u^* is given by

$$u^* = \frac{(\beta_{10} + \gamma_{10}) + \tilde{z}}{\beta_{12} + \gamma_{12}} \qquad (6.45)$$

which is obtained by substituting Equation 6.44 into Equation 6.43 and setting $\Delta^2 p = 0$. If aggregate demand is reduced below the level required to sustain u^*, we move to a point such as B at unemployment $u > u^*$ which is consistent with

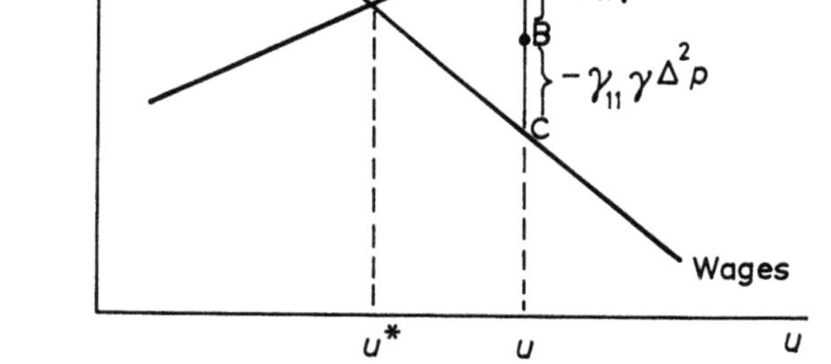

6.3 Equilibrium unemployment in a wage/price model

price and wage setting because inflation is falling. It is clear that the rate at which inflation falls depends on the vertical distance DC since this distance is equal to $(\beta_{11}+\gamma_{11})\gamma|\Delta^2 p|$. The precise point which emerges depends on the relative sizes of β_{11} and γ_{11}. However, this is not our main concern. The key point to note is that for any level of unemployment $u > u^*$, the reduction in inflation depends on the vertical distance between the price and wage lines at that point.

In the light of this framework, there are two particular aspects of the form of the $g(\cdot)$ function which are very important. These are first, its possible concavity and second, its dynamic structure. Taking concavity first, suppose that g has the form

$$g(u, u_{-1}, \Delta l) = \bar{g}(u), \quad \bar{g}' > 0, \quad \bar{g}'' < 0. \tag{6.46}$$

How and why this might arise is discussed at some length in Nickell (1986). There are, in fact, good arguments in favour of this functional form which have been recognized ever since Lipsey's original discussion of the Phillips curve in which he produced a simple aggregation argument to justify the curvature (see Lipsey, 1960). The implications of concavity are best seen diagrammatically and so, in Fig. 6.4, we present the wage—price lines based on Equations 6.39 and 6.40 with 6.46. When wage pressure is low ($\tilde{z} = \tilde{z}_1$), we have equilibrium unemployment u_1^*. If unemployment is actually at $u_1 > u_1^*$, then inflation falls at a rate proportional to D_1C_1. At a higher level of wage pressure, we have a higher natural rate u_2^* and when unemployment is at $u_2 > u_2^*$, then inflation falls at a rate porportional to D_2C_2. However, it is clear from this diagram that when $u_1 - u_1^* = u_2 - u_2^*$ then

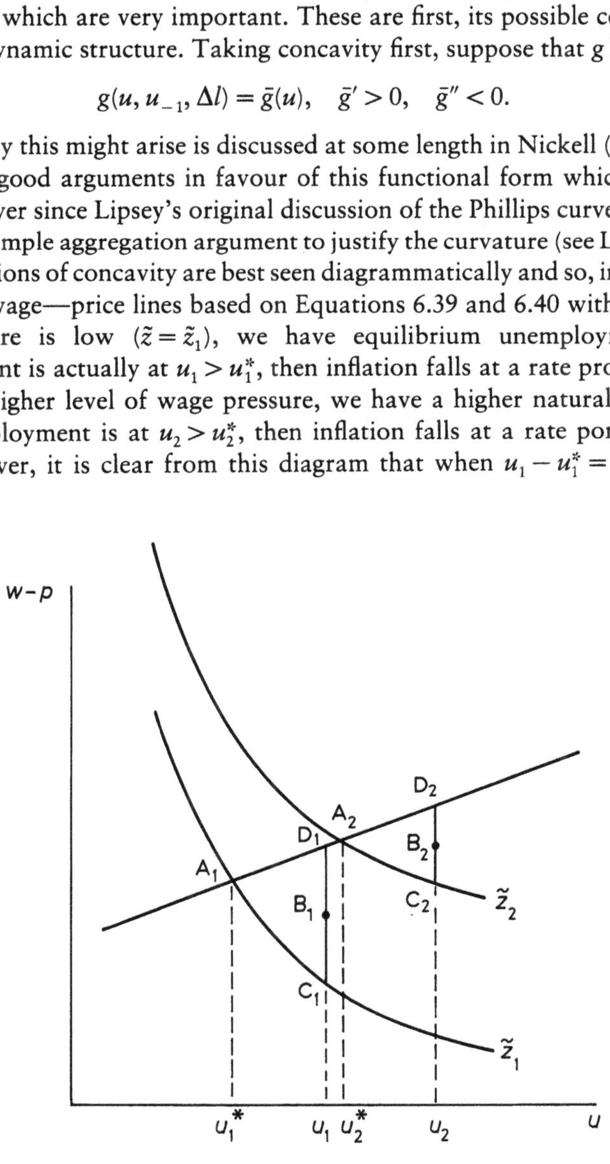

Fig. 6.4 The wage/price model with a concave unemployment effect on wages

$D_2C_2 < D_1C_1$ because of the curvature of the wage line. Thus, the concave shape of g means that as the natural rate gets higher and higher, excess unemployment becomes less and less effective at reducing inflation.

Turning next to the dynamic structure of g, suppose it has the form

$$g(u, u_{-1}, \Delta l) = \gamma_{12}u + \gamma'_{12}\Delta u, \quad \gamma_{12}, \gamma'_{12} > 0 \tag{6.47}$$

The theory underlying this structure we may leave until later although we have already seen how a positive lagged u effect may arise in our derivation of Equation 6.24. Using Equation 6.47, the basic equation (Equation 6.43) may be written

$$(\beta_{12} + \gamma_{12})u = (\beta_{10} + \gamma_{10}) - (\beta_{11} + \gamma_{11})\gamma\Delta^2 p + \tilde{z} - \gamma'_{12}\Delta u \tag{6.48}$$

So although the 'long-run' natural rate, u^*, is given by Equation 6.45, when unemployment is falling ($\Delta u < 0$), the effective natural rate appears to be somewhat higher since the term $-\gamma'_{12}\Delta u$ appears on the right-hand side. If we rearrange Equation 6.48 and make use of Equation 6.45 we obtain

$$\Delta^2 p = -\frac{[(\beta_{12} + \gamma_{12})(u - u^*) + \gamma'_{12}\Delta u]}{(\beta_{11} + \gamma_{11})\gamma} \tag{6.49}$$

From this we see that even if u is greater than u^*, any attempt to bring down unemployment will generate an increase in inflation if

$$-\Delta u > \frac{\beta_{12} + \gamma_{12}}{\gamma'_{12}}(u - u^*) \tag{6.50}$$

So, for example, if $(\beta_{12} + \gamma_{12})/\gamma'_{12}$ is 0.2, then even when unemployment is 4 points higher than the natural rate, an attempt to reduce employment at a rate of 1 point per year will generate increasing inflation throughout the four years it takes to reach equilibrium. So the existence of Δu terms in the wage equation clearly makes the reduction of unemployment that much more difficult even when starting from a position well above the natural rate.

This completes our discussion of the workings of the model and we now consider some of the results obtained when models of this type are confronted with the data.

6.5 EMPIRICAL ANALYSIS OF THE BRITISH LABOUR MARKET

As we have already noted, the key equations for our labour market model are those referring to employment, prices, wages and trade balance (6.26, 6.27, 6.28, 6.37 respectively). Estimates of these are presented in Table 6.1 and are taken from Layard and Nickell (1986). There are several points worth noting when comparing these with the theoretical model. First the price surprise or nominal inertia terms are captured by the second difference of the wage in the price equation and do not appear at all in the wage equation. This is common when using annual data (see Minford, 1983, for example) and need not concern us unduly since the parameters

Table 6.1 Estimates of a labour market model (1954–83)

Employment: $n-k = \underset{(4.9)}{2.57} + \underset{(8.2)}{1.06}(n_{-1}-k) - \underset{(2.6)}{0.36}(n_{-2}-k) + \underset{(3.2)}{0.067}\sigma - \underset{(4.9)}{0.285}(w-p)_{-1}$

SE $= 0.0088$, LM (autocorrelation, $\chi^2(2)$) $= 2.2$, Parameter stability, $\chi^2(7) = 5.1$

Prices: $p-w = -\underset{(4.2)}{4.18} + \underset{(5.0)}{0.54}(p-w)_{-1} + \underset{(2.1)}{0.038}\sigma - \underset{(4.2)}{0.34}\Delta^2 w - \underset{(3.8)}{0.24}\Delta^2 w_{-1}$

$-0.49(k-l)$

SE $= 0.015$, LM (autocorrelation, $\chi^2(2)$) $= 5.4$, Parameter stability, $\chi^2(5) = 1.2$

Wages: $w-p = \underset{(9.0)}{8.41} + 1.07(k-l) - \underset{(4.4)}{0.062}\log u + \underset{(2.5)}{0.50}c + \underset{(2.0)}{0.42}\Delta c + z$

SE $= 0.0145$, LM (autocorrelation, $\chi^2(2)$) $= 3.7$, Parameter stability, $\chi^2(7) = 0.9$

Balance of trade: $B/Y^{\mathrm{P}} = 41.68 + 361.0c_{-1} + 135.8v(p_{\mathrm{m}}-p)_{-1} - 39.8\sigma_{-1} + 24.7\mathrm{OIL}$

SE $= 10.46$, LM (autocorrelation, $\chi^2(2)$) $= 4.1$. Parameter stability, $\chi^2(4) = 4.4$

(i) $\sigma = (p^* - \bar{p}) + 10.76\mathrm{AD} + 1.028\mathrm{WT}$
$z = 0.039\,\mathrm{mm} + 0.18\rho + 0.50v(p_{\mathrm{m}} - p^*) + 0.42\Delta v(p_{\mathrm{m}} - p^*) + 0.085\mathrm{Up} + 0.18t_1 - 0.021\mathrm{IPD}$
n = employment, k = capital stock, σ = deviations of demand from full utilization output, w = hourly earning and employers' tax rate, p = GDP deflator, l = labour force, u = unemployment rate (so $u = 1\ 1 - n$), $c = v(p^* - \bar{p})$ = competitiveness, p^* = world price of manufacturing exports *in pounds*, $\bar{p}$ = TFE deflator at factor cost, v = share of imports in GDP, B/Y^{P} = (value of exports − value of imports)/nominal potential GDP. OIL = real value of North Sea oil production in terms of output prices, AD = adjusted budget deficit/potential GDP. WT = deviation of world trade from trend, mm = mismatch index, ρ = benefit replacement ratio, UP = union/non-union wage mark-up (a union power measure), t_1 = employers' labour tax rate, IPD = incomes policy dummy (1 in 1976–7, 0 elsewhere). A complete discussion of all these variables is contained in the Data Appendix of Layard and Nickell (1986).
(ii) t ratios in parentheses. Equations are estimated by non-linear 3SLS.

are consistently estimated despite the appearance of the price surprise in the error term. It is easier to detect the price surprise in quarterly data and some idea of the extent of nominal inertia can be gleaned from the quarterly wage equation reported in Layard and Nickell (1986, table 14). Second, technical progress makes no appearance essentially because its estimated coefficient is so close to zero that it was dropped. Given that the wage elasticity of employment is very close to unity, this accords closely with our prior expectations given that technical progress is labour-augmenting (note the restriction implied by Equation 6.18). Third, σ, the level of demand relative to full utilization output is a weighted sum of competitiveness, the adjusted budget deficit and the deviations of world trade from trend. The weights are estimated along with the remaining parameters of the model. Finally, the wage pressure variables include measures of mismatch, union power and the replacement ratio as well as some elements of the 'wedge', namely employers' labour taxes and the real price of imports. The latter has been split into competitiveness and the world 'terms of trade' between Britain's imports and world manufacturing output along the lines discussed in the previous section.

On the basis of these estimates, we may derive the (static) empirical equivalents to the price and wage equations (6.39, 6.40) which served as the core of our discussion in the previous section. These are

$$\text{Prices:} \quad p - w = \beta_{10} - \beta_{11}(p - p^e) - 0.279u - 1.07(k - l) \tag{6.51}$$

$$\text{Wages:} \quad w - p = \gamma_{10} - \gamma_{11}(p - p^e) - 0.327u - 0.081 \log u + 1.07(k - l) + \tilde{z} \tag{6.52}$$

where

$$\begin{aligned}\tilde{z} = {} & 0.051\text{mm} + 0.235\rho + 0.11\text{Up} + 0.41v(p_m - p^*) + 0.65(c - c^*) \\ & + 0.34\text{OIL} + 0.34\Delta v(p_m - p) + 0.55\Delta(c - c^*) - 0.037\Delta\text{OIL}\end{aligned} \tag{6.53}$$

and the variables are defined in the footnote to Table 6.1. For the moment we have left the nominal inertia terms unspecified. Furthermore we have removed all the lags, so these are the 'long run' solutions. To look at the implications for unemployment, we eliminate the real wage and setting $p - p^e = \gamma\Delta^2 p$, we obtain

$$0.606u + 0.081 \log u = (\beta_{10} + \gamma_{10}) - (\beta_{11} + \gamma_{11})\gamma\Delta^2 p + \tilde{z} \tag{6.54}$$

If we set $\Delta^2 p = 0$ and $c = c^*$, we may use Equations 6.54 and 6.53 to analyse changes in the natural rate, u^* (conditional on trade balance). In Table 6.2 we provide a breakdown of shifts in the natural rate since the 1950s. These and many related results are discussed at length in Layard and Nickell (1986) and are included here to serve as an illustration of what can be done within this kind of framework. The numbers themselves are of some interest, in particular, the strong *negative* impact of North Sea oil on the natural rate. This arises from the fact that North Sea oil generates a higher real exchange rate and hence lower wage pressure for any given level of demand and thereby enables the economy to be run at a higher level of activity. Of course, in order to do so, government policy must be used to offset the demand effects implicit in the higher real exchange rate.

Turning now to the unemployment inflation tradeoff, Equation 6.54 may be rewritten as

$$\Delta^2 p = -\frac{1}{\gamma(\beta_{11} + \gamma_{11})}[0.606u + 0.081 \log u - (\beta_{10} + \gamma_{10} + \tilde{z})] \tag{6.55}$$

From the definition of the natural rate, u^*, we can rewrite Equation 6.55 as

$$\Delta^2 p = -\frac{1}{\gamma(\beta_{11} + \gamma_{11})}[0.606(u - u^*) + 0.081(\log u - \log u^*)]$$

or

$$\Delta^2 p = -\frac{1}{\gamma(\beta_{11} + \gamma_{11})}\left(0.606 + \frac{0.081}{\bar{u}}\right)(u - u^*) \tag{6.56}$$

Table 6.2 Estimates of the natural rate of unemployment, Males, 1956–83 (percentages)

	1956–66		*1967–74*		*1975–79*		*1980–83*
Actual unemployment rate	1.96		3.78		6.79		13.79
'Natural' unemployment rate	1.96		4.19		7.63		9.07
Changes in the 'natural' rate		2.23		3.44		1.44	
Causes of the changes:							
Employers' labour taxes (t_1)		0.29		0.51		0.69	
Benefit replacement ratio (ρ)		0.64		−0.12		−0.15	
Unions (Up)		1.40		1.58		1.25	
Oil production (OIL)		—		−0.32		−1.73	
UK import/world manufactures price ratio ($p_m - p^*$)		−0.29		2.02		−0.17	
Mismatch (mm)		0.19		0.27		0.77	
Income policy (IPD)		—		−0.50		0.78	

(i) These numbers are taken from Tables 10 and 11 in Layard and Nickell (1986).

where $\bar{u}$ lies between u, u^* (from the mean value theorem). In order to see the implications of the concavity of $\log u$, note that $0.606 + (0.081/\bar{u})$ is 2.23 ($\bar{u} = 0.05$), 1.76 ($\bar{u} = 0.07$), 1.42 ($\bar{u} = 0.10$), 1.15 ($\bar{u} = 0.15$), 1.06 ($\bar{u} = 0.17$). Thus when (male) unemployment is at its current level (17%), the addition of an extra percentage point has less than half the inflation reducing effect as it would have if we had the 5 percent unemployment of the early seventies.

If we are to have some idea as to the actual order of magnitude of the inflation reduction we require some estimate of the nominal inertia effect from the wage equation (γ_{11}). If we use the estimated effect from the quarterly equation in Layard and Nickell (1986) and suppose that $\Delta^2 p$ and $\Delta^2 w$ are the same, then we estimate[9] $\gamma(\beta_{11} + \gamma_{11}) = 1.43$. So, at current levels of unemployment, every percentage point above the natural rate should reduce inflation by around 0.7 percentage points per annum. Although the numbers do not seem unreasonable, there is clearly a problem in explaining the facts since 1983. By our estimates, unemployment is more than 5 percentage points above the natural rate yet core inflation in Britain is hardly falling. Of course, the numbers we have presented do not take account of all the lags in the system and once these have worked through, they may accord more directly with the facts. However, the current state of affairs certainly provides a bit of a challenge unless one is prepared to allow an enormous rise in the natural rate in the recent past, which seems highly unlikely.

This challenge has generated a large number of suggested explanations, nearly all of which are concerned with the ways in which the level of activity in the labour market influences wages. At one extreme is the view that it doesn't (see Beckerman and Jenkinson, 1986 for example)[10] which seems to be somewhat at variance with the results presented here and, perhaps more comprehensively, in Grubb (1986) which covers the whole OECD and explains the Beckerman and Jenkinson results (see Grubb, p. 65). Somewhat less extreme is the view that wage determination resides in the hands of 'insiders' (Blanchard and Summers, 1986, for example). Alternatively, various compositional effects have been adduced so that it is only a certain sub-group of the unemployed who have any impact on wage determination. In Nickell (1987), a large number of different hypotheses are investigated and one that appears to be consistent with the data is the notion that individuals who have been unemployed for a long time exert little or no

[9] In order to generate the nominal inertia in the wage equation, we use the regression in note (iv) of Table 14 to compute the long-run relationship between $p - p^e$ and $\Delta^2 p$, noting that we must transform from quarterly to annual rates of change.

[10] Somewhat curiously, those who espouse the view that the level of activity in the labour market has no effect on wages typically argue that this implies that there is no natural rate. A glance at Equation 6.45 immediately reveals that this is not the case. If $\gamma_{12} = 0$, there is no unemployment effect on wages. This has no implications whatever for the existence or otherwise of the natural rate unless it should also happen that $\beta_{12} = 0$. In this case, of course, there is no long run equilibrium in the economy because the price setting behaviour of firms and wage determination in the labour market are inconsistent (in the absence of surprises) at any level of economic activity. This is akin to a market model in which the supply and demand curves are both horizontal at different levels.

downward pressure on wages. This would arise both because the long-term unemployed are less active in seeking work (see Jackman and Williams, 1985, for example) and because they become less desirable to employers. This, then, suggests that the impact of unemployment on wages will become attenuated as the proportion who have been without work for a year, say, increases. If we take the wage equation in Table 6.1 and add in the proportion of those unemployed over a year ($u52/u$) we obtain

$$\begin{aligned} w - p = {} & \underset{(6.3)}{5.59} + 1.07(k - l) - \underset{(7.8)}{0.104}\log u + \underset{(3.7)}{0.212}(u52/u) \\ & + \underset{(3.1)}{0.49}c + \underset{(1.5)}{0.29}\Delta c + z \end{aligned} \tag{6.57}$$

1956–83, se = 0.0114, LM (autocorrelation $\chi^2(2)$) = 1.27,
Parameter stability $\chi^2(10) = 0.53$ (See Nickell (1987, Table 1) for full details.)

The additional term clearly makes an important contribution to the equation and, in fact, has profound implications particularly for the short-term behaviour of the model. These arise because of the dynamic relationship between the proportion of long-term unemployed and the level of unemployment itself. This relationship is illustrated by the regression

$$\begin{aligned} u52/u = {} & \underset{(2.1)}{0.054} + \underset{(3.7)}{0.61}(u52/u)_{-1} - \underset{(5.6)}{2.41}u + \underset{(6.5)}{5.58}u_{-1} \\ & - \underset{(2.4)}{2.18}u_{-2} \end{aligned} \tag{6.58}$$

OLS estimation, 1956–83, se = 0.023, $\bar{R}^2 = 0.84$.

which makes good sense. As unemployment rises, the long-term unemployed proportion falls initially since, historically, increases in unemployment come about because of temporary rises in the inflow. In the long run, however, the long-term proportion tends to rise with unemployment. If we now use Equation 6.58 along with 6.57 we can generate a wage equation with some rather complex unemployment dynamics. The upshot of this is to produce the equivalent of Equation 6.54 which has the form[11]

$$0.104 \log u - 0.279u + 1.10\Delta u = (\beta_{10} + \gamma_{10}) - (\beta_{11} + \gamma_{11})\gamma\Delta^2 p + z \tag{6.59}$$

So the unemployment inflation tradeoff now becomes

$$\Delta^2 p = -\left[\left(\frac{0.104}{\bar{u}} - 0.279\right)(u - u^*) + 1.10\Delta u\right] \Big/ (\beta_{11} + \gamma_{11})\gamma \tag{6.60}$$

The addition of the duration composition term has both accentuated the concavity of the labour market activity effect on wages and induced complex dynamics. The former effect is illustrated by the fact that $(0.104/\bar{u}) - 0.279$ is 1.79 ($\bar{u} = 0.05$),

[11] Unlike Equation 6.54, this model is based on constant competitiveness (c = constant), not on constant deviations of competitiveness from its long-run level ($c - c^*$ = constant).

1.20 ($\bar{u}=0.07$), 0.76 ($\bar{u}=0.10$), 0.41 ($\bar{u}=0.15$), 0.33 ($\bar{u}=0.17$). So when (male) unemployment is at its current level (17%), the addition of an extra percentage point has around one quarter of the inflation reducing effect as the same addition on a 7% baseline, the 1979 level of unemployment. The reason for this is that as unemployment grows, the proportion of long-term unemployed also grows thereby continuously weakening the inflation reducing power of the unemployment. Furthermore, for this model we estimate $(\beta_{11}+\gamma_{11})\gamma$ to be 1.48 and so every percentage point on unemployment at a 17% baseline reduces inflation by a mere 0.22 percentage points per annum. So even if we are 6 points above the natural rate, given that unemployment is so high, the actual downward wage pressure induced by this slump is only enough to reduce inflation by 1.3 points per annum. Finally, the presence of the large Δu term in Equation 6.60 indicates that if we are 6 points above the natural rate, any attempt to reduce unemployment by more than around $1\frac{3}{4}$ percentage points per year is liable to lead to an immediate increase in inflation (the exact cutoff point is $6/(1.1/0.33)=1.8$). The reason for this is that as unemployment falls, there is a temporary shortage of active and desirable employees available in the unemployed pool as the short-term group are all snapped up. This leads to temporary inflationary pressure until everything unwinds.

To summarize, therefore, the role of the duration composition effect is to weaken the inflation-reducing effects of labour market slack and to make it appear that unemployment is much closer to the natural rate than it really is.

6.6 SUMMARY AND CONCLUSIONS

We have set out a model of the economy with price-setting firms and non-competitive wage determination. Despite this, the model has the natural rate property with all its attendant implications. We have also presented some of the consequences of confronting the model with aggregate data from the UK. This has enabled us to investigate the rise in unemployment in the last two decades and the unemployment–inflation tradeoff. One final point is worth making. The empirical results reported here should be viewed as a rough and ready calibration of an aggregate model, not as providing decisive quantitative insights into the inner workings of the labour market. The idea that we can capture all the nuances of wage determination, for example, in a single aggregate equation is clearly ludicrous. The results presented here should, therefore, be treated with caution.

REFERENCES

Beckerman, W. and Jenkinson, T. (1986) What stopped the inflation? Unemployment or commodity prices? *Economic Journal*, **96**, 39–54.

Bils, M. (1985) Empirical structural evidence on wages, prices and employment in the US, MIT (mimeo).

Blanchard, O. and Summers, L. (1986) Hysteresis and the European unemployment problem, *NBER Macroeconomics Annual* (ed. S. Fischer).

Carlin, W. and Soskice, D. (1985) Real wages, unemployment, international competitiveness and inflation; a framework for analysing closed and open economies, University College, Oxford (mimeo).

Cowling, K. and Waterson, M. (1976) Price–cost margins and market structure, *Economica*, **43**.

Dixon, H. (1986a) Unions, oligopoly and macroeconomic policy: a simple two-sector model, Birkbeck College (mimeo).

Dixon, H. (1986b) A simple model of imperfect competition with Walrasian features, Birkbeck College, economics discussion paper 186; *Oxford Economic Papers* (forthcoming).

Grubb, D. (1986) Topics in the OECD Phillips curve, *Economic Journal*, **96**, 55–79.

Hart, O. (1983) A model of imperfect competition with Keynesian features, *Quarterly Journal of Economics*, **97**, 109–38.

Jackman, R. and Williams, C. (1985) Job applications by unemployed men, Centre for Labour Economics, London School of Economics, Working paper 792.

Johnson, G. and Layard, R. (1986) The natural rate of unemployment: explanation and policy, in *Handbook of Labor Economics* (eds O. Ashenfelter and R. Layard), North–Holland, Amsterdam.

Layard, R. and Nickell, S. (1986) Unemployment in Britain, *Economica* (Special Issue on Unemployment), August, S121–69.

Lipsey, R. G. (1960) The relation between unemployment and the rate of change of money wage rates in the United Kingdom 1862–1957: a further analysis, *Economica*, **27**, 1–31.

Minford, P. (1983) Labour market equilibrium in an open economy, *Oxford Economic Papers*, **35** (supplement), 207–44.

Mirman, L. J., Tauman, Y. and Zang, I. (1986) Ramsey prices, average cost prices and price sustainability, *International Journal of Industrial Organization*, June.

Nickell, S. J. (1987) Why is wage inflation in Britain so high? *Oxford Bulletin of Economics and Statistics* February.

Rotemberg, J. J. and Saloner, G. (1986) A supergame-theoretic model of price wars during booms, *American Economic Review*, June, **46**, 390–407.

Stiglitz, J. E. (1986) Theories of wage rigidity, in *Keynes' Economic Legacy* (eds J. L. Butkiewicz, K. J. Koford and J. B. Miller), Praeger, New York.

Wren-Lewis, S. (1985) Imperfect competition and effective demand in the long-run, National Institute of Economic and Social Research, London, discussion paper 98.

7

An international perspective

MARTYN ANDREWS

7.1 INTRODUCTION

By way of contrast with the rest of this book, this chapter presents a partial survey of the methodologies used in the modelling of the aggregate labour markets of non-UK economies, albeit from a UK perspective.[1] The first of two objectives of this exercise is to examine how representative modelling practice in the UK is, that is whether any of the seven paradigms outlined in Chapter 2 are used more extensively outside the UK, or whether there are some that are primarily employed by UK investigators, nothwithstanding the obvious point that Chapter 2 does not claim to present a *comprehensive* survey. A second objective is to critically examine the fairly widespread research activity where a *number* of economies are the focus of a particular study. It appears that there are two quite distinct schools of thought as to the usefulness and objectives of these exercises, and so this particular controversy is aired up-front in Section 7.2.

The size of a study of this kind is almost unbounded, and in applying constraints Occam's razor is wielded across the following dimensions. We shall survey models where at least *two* of the following four aggregates are endogenously determined, namely wages, employment, unemployment, and the labour force, therefore excluding analysis of comparative work across countries based on just *one* equation. Moreover we restrict attention only to models that are estimated using time-series data. Thirdly we focus on the member countries of the OECD, since they are broadly comparable 'market' economies.[2]

[1] A touch of native colouring is almost an unavoidable consequence in a study of this kind. As will become apparent this chapter owes a substantial debt to the proceedings of a recent conference published in a Special Issue of *Economica*, on unemployment. Organized by the Centre for Labour Economics and held in Sussex, England, in May 1985, its purpose was to examine and explain "one of the most remarkable features of recent economic history... [namely] the remorseless rise in unemployment throughout the industrialized countries" (Bean *et al.*, 1986, p. 1).

[2] The Organisation for Economic Cooperation and Development (OECD) comprises 24 market economies: Australia (A), Austria (Au), Belgium (B), Canada (C), Denmark (D), Finland (F), France (Fr), Germany (WG), Greece (G), Iceland (Ic), Ireland (Ir), Italy (It), Japan (J), Luxembourg (L), The Netherlands (H), New Zealand (NZ), Norway (N), Portugal (P), Spain (Sp), Sweden (Sw), Switzerland (Ch), Turkey (T), the United Kingdom (UK), and the United States (US).

The main substance of this chapter is presented in Section 7.3, which cross-references particular countries with particular approaches to modelling the labour market aggregates. In order to incorporate a structure, the reader is invited to visualize a matrix whose columns refer to countries and whose rows refer to model-types. In particular the model-types are precisely those outlined in Chapter 2, which for convenience we set down once more:

The standard model (A) and the wage-gap model (D);
The market-clearing model (B);
Disequilibrium: single market, model (C);
Disequilibrium: multimarket, model (E);
Imperfect competition (F); and
Non-structural models (G)

For the purposes of this exercise the standard model (A) and the wage-gap model (D), being quite similar, are placed together on one row. In order to reduce the number of columns, the twenty-four OECD countries are grouped as follows:

North America (US, Canada);
Non-UK European;
Other OECD (e.g. Australia and Japan);
Miscellaneous; and
'All'

The above matrix has 30 cells, each of which represents many different studies. In order to apply constraints attention is focused only on studies that have either a distinguishing characteristic, or are representative of a particular model-type or group of economies. The reader is also advised that although there will be some blanks cells, this does not imply that a particular model-type has never been applied to a particular group of countries. Elements in the 'miscellaneous' category refer to the modelling exercises of organizations not affiliated to one specific country, such as the IMF, the World Bank, the OECD, and the EEC; and models of many countries supported by a particular organization such as the Federal Reserve Board. The final column is the most interesting, representing the research of one particular group of economists, usually of the same nationality, whereby equations of the same basic specification are estimated for a *number* of economies. This we refer to as 'multi-country studies'.

7.2 MULTI-COUNTRY STUDIES: SOME COMMENTS

As already emphasized the main focus of this chapter is methodology. Thus we are not *directly* concerned with studies that attempt to provide explanations of a fairly common cross-country phenomenon such as the rise in OECD unemployment; nor those which seek to establish the degree of so-called nominal wage rigidity for different countries. Rather we seek to discover whether the same paradigms and general methodological approaches are used in analysing these types of problem, or whether there exist different approaches depending on the country or countries of interest. For issues that are broadly applicable across a comparable set of countries, such as OECD unemployment, we quote Solow in considering the usefulness of an intercountry perspective:

> *"We can all hope to learn something from cross-country comparisons. One of the few good ways we have to test analytical ideas is to see whether they can make sense of international differences in outcomes by appealing to international differences in institutional structure and historical environment. The right place to start is within each country separately, studied by someone who knows the peculiarities of its history and its data.*
>
> *You might think that this ought to be too obvious. But in fact the usual approach is just the opposite. More often than not we fail to take institutional differences seriously. One model is supposed to apply everywhere and always. Each country is just a point on a cross-section regression, or one among several essentially identical regressions, leaving only grumblers to worry about what is exogenous and what is endogenous, and whether simple parameterizations do justice to real differences in the way the economic mechanism functions in one place or another,"*
>
> (Solow, 1986, p. 23)

Solow implicitly highlights two separate complaints, namely that when examining many issues an international perspective is not always considered, and even when it is, the methodology may be suspect. Gordon concurs with both points when examining US wage and employment flexibility, noting that it "seems remarkable that the modern American literature on labour market contracts contains no mention of cross-country differences in the extent of wage flexibility, much less any explanation of these differences" and "any economist who dares mention institutions as central determinants of macroeconomic performance had better tread carefully, lest he be branded a dangerous renegade or traitor" (Gordon, 1982, p. 14).

This chapter is organized with Solow's advice in mind. When cross-referencing countries by model-types, given there *are* inter-country differences in custom, institutional arrangements, environment, and history, we seek to establish whether these are reflected in the way the labour market is modelled. Emphasis is given to those single-country studies which take what we shall call 'background' into account in the specification and interpretation of the formal models estimated.

Our concern in the remainder of this sub-section is with investigators whose "usual approach is just the opposite", in Solow's words. The essential characteristic here is that the same basic specification is estimated with data for a number of different countries. Often this is done with an explicit comparative objective in mind at the outset, or sometimes as the second stage of a research program where the investigator, having successfully developed an econometric model relating to his own country, wishes to discover whether it is of more general applicability by extending it to a number of similar economies.

The main criticism of this approach arises because, in the development of formal, or econometric, models, the cross-country differences relating to institutional arrangements also need formalizing. Typically this is not the case, for two reasons. First a knowledge and understanding of the history and backgrounds of say between five and twenty other countries is too large a task for one or a small number of investigators, particularly if not natives of the countries in question. Second, the appropriate data often are not available. The corollary of these arguments forms the basis for the material discussed in Section 7.5, namely that comparative studies containing less formal or more descriptive analysis might be more appropriate; and that only a small number of countries be considered. However, methodology of this kind is extremely rare, compared with the type of multi-country studies surveyed in Section 7.4. These have many advocates. Newell and Symons, in estimating wage equations/Phillips curves for 16 OECD countries, argue as follows:

> *"This approach effectively eliminates one source of disparate results: tendentious specification-search with single-country data. Indeed the multi-country approach allows one to test different models in an open-ended way that would be unconvincing with the data of a single country."*
>
> (Newell and Symons, 1986, p. 2)

The fundamental problem being alluded to is what is commonly known as 'data-mining', one description being:

> *"A systematic search, varying the lag structure, variables included, and other features in order to find the best-fitting equation, artificially reduces the equation standard error and risks selecting an equation where the variables are 'significant' by chance rather through any real causal connection. This danger is greatest in single-country studies where the numbers of variants can approach, or even exceed, the degrees of freedom in a regression. Correlations found consistently in the data of many countries are less likely to be purely chance occurrences."*
>
> (Grubb, 1986, p. 57)

Thus the two schools of thought almost differ as a matter of principle. Some investigators believe that a hypothesis or regression is valid *only* if it has support from other similar countries, usually to counter criticisms of data-mining; whereas others believe that no two countries are the same, even the US and Canada. Can the approaches be reconciled? Assuming there *are* inter-country differences in 'background', the answer depends on what the implications are in imposing the same regression structure throughout, where *only* variation in parameter estimates can account for inter-country differences in background. Unless the omitted effects vary over time in the same way for each individual country, we should not expect to find the same correlation structure between endogenous variables, although the use of the same regression model implies otherwise. Of course, as with all models that incorrectly impose a false null hypothesis, useful *diagnostic* information is then generated, for example 'interesting' or 'implausible' parameter estimates for particular countries, or groups of countries. This helps identify those that are broadly similar in 'background', and also leads to well-known characterizations such as the US exhibits nominal wage rigidity in contrast to European countries who have real wage rigidity. Indeed the advocates of this type of exercise are aware of the pitfalls; Newell and Symons, continuing from above, note

> *"Nevertheless the approach has the added complication that one has to distinguish cases where sporadic results are due to chance from cases where such results are due to different properties of markets. This is usually quite difficult.*
>
> (Newell and Symons, 1986, p. 2)

which in principle goes some way towards reconciling the differences in approach noted by Solow.

The above discussion forms the basis of the distinction made between the studies examined in Sections 7.4 and 7.5, in that the latter group attempt to analyse, or correlate, the variation in parameter estimates across country in terms of known and observed differences in so-called background.

7.3 SELECTED APPLICATIONS OF LABOUR MARKET MODELS

In this section we attempt to cross-reference the model-types discussed in Chapter 2 with non-UK economies. This section is concerned with single country applications, bearing in mind Solow's advice that the right place to start is within each country separately. Thus this chapter can be thought of as corresponding to the main body of the matrix alluded to in Section 7.1.

7.3.1 The standard model

The top row of our matrix is a natural starting place, since applications of the standard model (A) are pandemic. Nowhere is this more apparent in the treatment of employment, unemployment and wages in the large number of large-scale macroeconometric models of each OECD country. The general proposition here is that almost all the labour markets of the large-scale macroeconometrics models follow (A), simply because, being the original labour-market model, macroeconometric modelling practice tends to lag behind current academic research. In this sub-section on 'on-model' practice, we first examine a selective number of macroeconometric models, to give a flavour of the standard approach. This is then distinguished from the more academic 'off-model' practice, to which the rest of this chapter is devoted.

For each country there exist a number of large-scale macroeconometric models, supported either publicly or privately, or both, but typically there is no organization that exists independently to initiate comparative work. For the UK the ESRC Macroeconomic Modelling Bureau is a clear exception, and in its first review, Wallis (ed.) *et al.* (1984) examined the labour markets of the six UK macroeconometric models.[3] This was in part a response to Nickell's caustic comment that "the notion of a labour market obviously lies somewhere in the background in these models but it seems to have moved so far back that it has disappeared from view" (Nickell, 1984, p. 13). The main finding was "that many of the model structures are not built around the concept of market, and labour demand, labour supply, and wages are often modelled in a way that leaves the underlying market structure ill-defined" (Wallis *et al.*, 1984, p. 93). Thus for the UK only three years ago the identification of a model structure with Equations 2.1, 2.9 or 2.10 was not obvious, particularly for the second wage equation/Phillips curve. More recently, however, matters have improved, most of the UK models conforming to model (A).[4]

The absence of a 'Bureau' in other countries does not imply a lack of useful survey material. Brooks (1984) systematically compares seven US macroeconometric models in an attempt to place them in the framework of standard classroom macro-paradigm. He describes the typical labour market as follows:[5]

> *"The demand for labour is determined by aggregate demand and the nature of the function governing the production of aggregate output. Thus GNP, derived by adding up final demands, together with the average productivity of the work force, determine total employment either in terms of hours or workers. This labor demand implies a certain degree of tightness or slack (typically measured by the unemployment rate) in labor markets depending on labor supply. Together with price expectations this labor-market tightness indicator determines the rate of wage inflation. The shape of this 'Phillips' curve is a matter of considerable research and differs from model to model. Most of the models provide a relationship that will allow some short-run expansion of labor demand without a sharp increase in wages, especially if there is sufficient slack in labor markets. While the longer-run properties of these equations are rarely obvious, there is a general sense that, if not exactly vertical in the long run, the typical 'Phillips' curve will provide for continued increases in inflation rates once the economy has*

[3] Namely those of Cambridge Growth Project (CGP), the City University Business School (CUBS), Liverpool University (LPL), the London Business School (LBS), the National Institute of Economics and Social Research (NIESR) and the Treasury model (HMT).

[4] Loosely speaking CGP, LBS, NIESR and HMT correspond to model (A); LPL to (B); and CUBS to (C). This classification is open to interpretation, however. For further details see Wallis (ed.) *et al.* (1984, 1986).

[5] These are the models of Data Resources, Inc.; Chase Econometric Associates, Inc.; Wharton Econometric Forecasting Associates; Townsend–Greenspan, Inc.; Citibank Information Services, Inc.; Federal Reserve Board (the MPS model); and the Bureau of Economic Analysis of the Commerce Department.

reached full capacity. There is no long-run, permanent trade-off between inflation and unemployment."

(Brooks, 1984, p. 5)

This he represents, in the notation of Chapter 2:

$$n = f_1(y, \lambda)$$
$$u^r \equiv (L - n)/L$$
$$L = f_2(\text{POP}, t)$$
$$\dot{w}/w = f_3(u, \dot{p}/p, \hat{\dot{p}}/p)$$
$$\hat{\dot{p}}/p = f_4(\dot{p}/p).$$

(λ is productivity.) f_1, f_2 and f_3 correspond to Equations 2.4, 2.10 and 2.9 respectively, and f_4 is some forecasting scheme such as adaptive expectations.

A similar approach is adopted in Canadian macroeconometric modelling. In a recent report prepared for the Economic Forecasting Division of the Department of Finance, by Grady and Associates Ltd (1985), eleven Canadian models are comprehensively surveyed.[6] Again all specify an accelerationist or expectations augmented Phillips curve, although the natural rate of unemployment is not defined in two. For most of the models inflation expectations are formed adaptively, although there are more "novel approaches" (Grady and Associates Ltd, 1985, p. 42). There is a unit coefficient $\dot{p}$ (or $\hat{\dot{p}}$) in two of the models. For labour demand interrelated factor demands systems have been specified, usually with a Cobb–Douglas technology. Labour supply is usually determined by underlying demographics and participation rate equations; and for one model in particular (SAM) an intertemporal optimizating decision underpins the households' labour supply decision, as in Lucas and Rapping (1969).

Thus in support of our proposition above, for the modelling of wages and employment in the large-scale macroeconometric models, there is considerable uniformity in the basic structure for those of the two North American economies. The story is the same the whole world over, the above discussion being sufficiently representative of most applications of model (A).[7]

[6] These are: CANDIDE, Economics Council of Canada; TIM, the Informetrica Model, Informetrica Ltd.; RDX2, Research Department Experimental Model, Research Department, Bank of Canada; CHASE Econometric Model, Chase Econometrics; DRI Model of the Canadian Economy, Data Resources Canada; FOCUS, Forecasting and User Simulation Model, Policy and Analysis Programme, Institute for Policy Analysis, University of Toronto; MTFM, Medium-Term Quarterly Forecasting and Simulation Model, Conference Board of Canada; QFS, Quarterly Forecasting and Simulation Model, Economic Forecasting Division, Department of Finance; RDXF, Research Department Experimental Forecasting Model, Research Department, Bank of Canada; MACE, University of British Columbia; and SAM, Small Annual Model, Research Department, Bank of Canada.

[7] It is impossible to give a comprehensive list of references that establish this proposition convincingly. If readers are interested in discovering how wages and employment are modelled in macroeconometric models around the world, they might start with researching the four volumes of *Economic Modelling*, not forgetting that organizations such as the OECD have models for more than one country, namely the INTERLINK model (Helliwell *et al.*, 1985) as does the EEC; with the

(*Contd. on next page*)

Turning now to 'off-model' applications of models (A) and (D), the most well-known application of (D) is that of Bruno and Sachs (1985), their summary of much research into the role of wage-gaps in explaining the stagflation of the late 1970s, which we examine later in Section 7.5. However, model (A) attracts much attention in current research. In the recent *Economica* 'unemployment' Special Issue, containing contributions for nine OECD countries, half were of model (A) type: Gregory (1986), for Australia; Pichelmann and Wagner (1986), Austria; Malinvaud (1986), France; Franz and Konig (1986), West Germany and Hamada and Kurosaka (1986), Japan. The last of these is of some interest, where some attention is paid to Japan's quite different 'background'; (see in particular section 3). Our next objective is to consider one column of our matrix, to see to what extent any of the other model-types, apart from (A), are considered in studies of the aggregate labour market; our choice being Australia.

7.3.2 Australian labour market models

In this section we briefly consider some recent studies of wage and employment behaviour for Australia. Although a fairly arbitrary choice this is not entirely so, since Australia's institutional arrangements and historical background are somewhat unique, and whose consideration are an essential requirement for successful empirical work, if Solow's earlier critique has any substance. For example, in his extremely comprehensive survey of the Australian labour market, Mitchell notes at the outset:

> *"Conditions in the Australian labor market are in some ways very familiar to a foreign observer, and in other ways very different... [He or she] is on least familiar ground in the realm of industrial relations and wage determination. Australia has a major union federation... to whose affiliates belong the bulk of the countries members. Its counterpart on the employment side... has a clear role as the most prominent, national management group. Yet at that point similarities with most other countries fade."*
>
> (Mitchell, 1984, p. 1)

A focal point for many studies of the Australian labour market is the high wage inflation and associated high real wage levels experienced in the mid-1970s, often thought to be central to explanations of the economy's failure to maintain full employment from about the same time, via a wage-gap argument. Of course the real wage itself also requires explanation. Equations 2.1 and 2.9 are a natural vehicle for empirical work, with particular attention paid to the degree to which the real wage influences labour demand, other important determinants of demand/

(Contd. from previous page)

EUROLINK (Bucher and Rossi, 1985) and COMET models; and the Federal Reserve Board with their MCM model (Board of Governors, 1984). Examples of other interesting and notable references are Eliasson (1986), who develops a micro-to-macro model for Sweden; van Wijnbergen (1985), who develops a model stressing the interdependencies between the OECD, OPEC and LDC blocks of countries; and so forth.

employment, and the degree to which the real wage responds to labour market conditions. A correctly specified model should then explain *both* the behaviour of wages and employment in the turbulent 1970s, while at the same time being consistent with the particular institutional arrangements peculiar to Australia.

Central to Australia's unique position is a system of compulsory arbitration of industrial disputes, which also sets minimum wages and other minimum conditions of employment for most occupations. The system is a quasi-judicial process consisting of mainly federal and state tribunals, the first of these forming what is known as 'The Commission'.[8] However, private bargaining takes place over and above these minimum requirements. Thus one school of thought argues that wage setting in Australia is centralized and heavily regulated.

The Commission's response to the wage inflation of the mid-1970s was to attempt to operate an incomes policy for the 90% of wage and salary earners under its direct or indirect jurisdiction, comprising a national wage system that (almost) fully indexed wages between April 1975 and June 1981. Similarly since 1983 there has been full wage indexation, preceded by a brief wage pause. Clearly the effectiveness of these incomes policies depends on the degree of which the Commission "merely acts as veil for market forces", a view in direct contrast to that above. This possibility motivates Gregory's (1986) recent attempt "to build up sufficient understanding of the wage determination process so that some evaluation can be made as to the effectiveness of these national wage policies" (Gregory, 1986, p. 53), given that the extensive literature is not conclusive on this hypothesis. The background history to empirical wage equations in Australia appears to convey a story common to other OECD countries, the supposed stable relationship between nominal wage changes, unemployment, and excess supply of labour, breaking down with the more turbulent data in the mid-1970s; and the hypothesis of a vertical Phillips curve becoming less and less difficult to reject. Of the current evidence some studies take account of the arbitration system explicitly, although most do not.[9]

Although Gregory's study contains a good account of the institutional background (from which much of this is taken), the regression evidence boils down to the estimation of a simple APC (augmented Phillips curve), Equation 2.9, for three different samples, so as to examine whether indexation policy kept nominal wage increases at lower levels than otherwise essentially by noting whether there is any within-sample predictive failure for the indexation period. A parallel equation is estimated where unemployment is replaced by the level of overtime hours, to distinguish between the pressure exerted on wages by the labour utilization rate within the firm, and the utilization rate of the whole economy, with some success. Gregory also notes that comparing the wage inflation history of Australia with the average of six other OECD countries reveals some

[8] The Australian Conciliation and Arbitration Commission.

[9] See the references cited in Gregory (1986, p. 57) and the other contributions in Caves and Krause (1984).

similarity, but again indicating that in the indexation period Australian wage inflation was lower than its competitors, a basis with which to test the efficacy of Australian wage policy. However, his main conclusions are quite straightforward, namely that regressions which exhibit poor fits, autocorrelation, and unstable coefficients, are not the basis for understanding wage determination in Australia, nor indeed anywhere else.

By contrast recent econometric work for labour demand is empirically more satisfactory, the most notable of which is Symons (1985a), based on his earlier work for the UK (Symons, 1981, 1985b). Although real wage had a negative impact on, or more correctly association with, demand/employment in Australia in the mid-70s, there are periods when the two variables covaried positively. This is consistent with a shock to the labour demand schedule. The increase in the price of raw materials and fuel that came of OPEC(2) – it is generally accepted that Australia was somewhat cushioned from OPEC(1) – implies that models of labour demand should take this variable into account, as emphasized by Bruno and Sachs (1985). Consequently Equation 2.7 is the appropriate specification, the key parameter estimates taking values typically found for other OECD countries.

The two equations discussed above together form model (A), although the labour demand equation of Symons is neoclassical in flavour since the authorities are impotent at shifting the schedule directly. The standard non-competitive specification is given by model (F), whose key equation (Equation 2.8′) is estimated by Trivedi and Alexander (1986). In looking for a role for competitiveness and some real aggregate demand proxies, they find that the real money supply is the only significant variable capable of supporting a Keynesian viewpoint. Thus it remains unresolved as to whether Equation 2.7 or 2.8′ is best supported by the data for Australia.

7.3.3 Non-structural evidence, as applied to the United States

In our examination of the modelling of wages and employment for Australia, it is clear that standard methods are usually applied in investigating the usual issues relating to market behaviour, namely whether wages are effectively set at competitive levels, and whether demand management can induce changes in employment via the labour demand schedule. We now consider what model-types are brought to bear in the modelling of labour market aggregates for the United States. As we shall see shortly, the neoclassical or equilibrium labour-market hypothesis attracts much attention in the US, as exemplified by Lucas and Rapping's (1969) competitive contribution, notwithstanding Rosen and Quandt's (1978) explicit attempt at modelling disequilibrium. (These two papers can be thought as the original entries for models (B) and (C) in the US column of our matrix.[10])

[10] The author is not aware of entries for models (E) and (F), although no doubt they exist. None the less disequilibrium modelling is not as popular in the US as it is in Europe; and the imperfectly competitive approach of Layard and Nickell is still very new.

Given that 'wage gap' type arguments are so popular in current labour/macro analysis, it is not obvious why testing whether the labour market clears has such a prominent role in the US, when viewed from an OECD context. However as soon as inter-country comparisons are brought to bear this is less puzzling as the US economy often turns out to be an outlier, and so there are more grounds for believing that an equilibrium structure is more appropriate, also given much anecdotal evidence suggestive of a free-enterprise culture. Of direct relevance to the wage-gap argument is the observation that unemployment spells tend to be less sustained in the US; and are in fact currently moving in the opposite direction to most other countries.[11] Strongly competitive forces might also explain why the US is characterised as having nominal, rather than real, wage rigidity, as is often thought to be the case.

We now consider one particular application of model (G) of Chapter 2, that of Altonji and Ashenfelter (1980), in examining the equilibrium hypothesis that underpins the labour market structure as given by model (B). This precludes the necessity of estimating the structure itself. However, as the authors themselves note that "a richer structural labour market model... [is also required as]... part of any fair test of the labour market equilibrium hypothesis" (Altonji and Ashenfelter, 1980, p. 240), but our concern here is with methodology, not with testing the hypothesis *per se*.

That the equilibrium hypothesis is important is not in doubt, simply because of the policy implications that follow. The question is whether it is consistent with the kind of facts already noted. Specifically the most important of these is that employment is known to move cyclically *and* with a strong negative relationship with the percentage of people who say they would like to work, but cannot. The Lucas and Rapping (1969) model provides a convincing reconciliation of this with the theory, incorporating a highly elastic short-run labour supply schedule, which is derived from the possibility that workers can easily substitute leisure and goods across time, having access to perfect capital markets. Altonji and Ashenfelter simply examine whether the observed time-series behaviour of wages and employment are actually consistent with the equilibrium hypothesis.

As noted briefly in Chapter 2, the Lucas and Rapping model posits that unemployment is positively related to the gap between the normal real (consumption) wage, W_t^*, and its current value, W_t (dropping the superscript c for convenience):

$$u_t^r = c_0 + c_1(W_t^* - W_t) \quad c_1 > 0 \tag{7.1}$$

Lucas and Rapping substitute out W_t^* using the adaptive expectations hypothesis. Assuming that the real wage is adequately represented by a simple autogressive process, AR(n), tantamount to setting $a_{21}(L) = 0$ in Equation 2.21,

$$[1 - a_{22}(L)](w - p)_t = d_0 + d_1 t \tag{7.2}$$

[11] Including Canada. Since the US and Canada have always been considered almost identical economies, this phenomenon merits attention (Ashenfelter and Card, 1986).

then the equilibrium hypothesis is in difficulty should the real wage follow a random walk, simply because then $W_t^* - W_t$ is constant, as is unemployment via Equation 7.1. Since unemployment is observed as having a cyclical component, this must then be attributable to something else apart from the assumed behaviour of suppliers of labour. This is most easily seen by supposing that W_t^* can be represented by ${}_t\hat{W}_{t+1}$ in the usual notation, collapsing the whole of the future into one period, noting that the optimal forecast from a random walk is the current value, W_t.

Testing for a random walk against a more general AR (n) alternative hypothesis typically follows a two-stage procedure. First Equation 7.2 is reparameterized so that the possible unit root is isolated, which is then tested. If accepted, that is $1 - a_{22}(L) = (1 - L)[1 - a'_{22}(L)]$, the model becomes a standard autogression in the *rate of change* of real wages,

$$[1 - a'_{22}(L)]\Delta(w - p)_t = d_0 + d_1 t.$$

If, on re-estimation, its autoregressive part is not significant, then the random walk hypothesis is not rejected. It should be noted that the econometrics is not straightforward, there being no asymptotic theory for non-stationary processes; also diagnostic checks on the residuals for possible autocorrelation are necessary at both stages, for obvious reasons.[12]

Using US and UK quarterly data on gross earnings deflated by the consumer price index, Altonji and Ashenfelter conclude that for the US there is little evidence against the random walk hypothesis, although the weaker root unit hypothesis is probably the best description of the data; but for the UK the random walk hypothesis should be rejected. For both countries the results are more favourable to the null hypothesis when the significant trend term is dropped. Not surprisingly, when forecasts generated from some appropriate version of the real wage process are included as regressors in the unemployment equation, they are generally not

[12] Writing $a_{22}(L)$ in Equation 7.2 as $a_1L + a_2L^2$, which is sufficiently general for illustrative purposes, we have:

$$(1 - a_1L - a_2L^2)(w - p)_t = \cdots$$

or equivalently

$$(1 - \lambda_1 L)(1 - \lambda_2 L)(w - p)_t = \cdots$$

where λ_1 and λ_2 are the two roots. The relationships between the parameters to be estimated, a_1 and a_2, and the roots, λ_1 and λ_2 are:

$$a_1 = \lambda_1 + \lambda_2 \quad a_2 = -\lambda_1\lambda_2$$

The first step is to test $H_0\colon \lambda_1 = 1$. Under H_0 $a_1 = 1 + \lambda_2$ and $a_2 = -\lambda_2$, which is equivalent to H_0: $a_1 + a_2 = 1$. The problem here is that the usual statistic does *not* have a t-distribution under H_0. If H_0 is accepted, the model reduces to

$$(1 + a_2L)\Delta(w - p)_t = \cdots$$

The random walk hypothesis is tested in the second stage by examining whether in a regression of $\Delta(w - p)_t$ on its first lag, this variable is significant.

significant, which can be thought of as testing $a_{12}(L) = 0$ in the top row of Equation 2.21.

Thus Altonji and Ashenfelter's results are not favourable to the Lucas and Rapping intertemporal substitution labour supply version of the equilibrium hypothesis. See also Altonji (1982) and for an excellent survey Kniesner and Goldsmith (1986). There are two further points. First, acceptance of the random walk hypothesis is quite supportive of an explicit disequilibrium structure, as it is the stochastic counterpart of fixed real wages. Second is the point that testing for real wage rigidity is not the same as testing for nominal wage rigidity, the latter supposed to be the appropriate description for the behaviour for wages in the US. A time-series description of the data could easily be estimated in the set-up of Equation 2.21, where *both* w and $w-p$ would be jointly modelled. However, typically, as noted below, it is the *ad hoc* Equation 2.9 that provides the framework, since nominal wage rigidity does not fit easily into models where agents are rational optimizers, notwithstanding the theoretical contributions along the staggered and implicit contract lines.

In concluding this section, it should be emphasized that whilst it is only the autoregressive process for wages that is examined explicitly, the implied structure is a *vector*-autoregressive; it just turns out that there is no interrelationship between (real) wages and employment. One example of the vector-autogressive approach to examining other non-competitive hypotheses underlying the US labour market models is Ashenfelter and Card (1982); another, set in a multi-country context, we turn to shortly.

7.3.4 The multimarket disequilibrium model, as applied to Belgium

We conclude our survey of single-country studies by examining the intersection of both a row and column of our matrix thus far neglected, taking us to models of type (E) as applied, quite commonly, to European economies. Our example is that of Sneessens and Drèze (1986), which is an attempt at estimating a coherent two-market rationing (or disequilibrium) model of the Belgian economy. It draws mainly on previous work by Sneessens, whose (1983) paper was used to illustrate this type of paradigm in Chapter 2.

The paper's central theme is Belgian unemployment, and descriptive evidence is provided as to why the distinction between classical and Keynesian unemployment might be important. If OECD unemployment has given cause for concern, then the rise in Belgian unemployment since 1974 "has been appalling" (Sneessens and Drèze, 1986, p. 89). The gap between Belgium's unemployment rate and a nine-country EEC average has risen steadily to almost four percentage points, and is related to the following facts. For the period 1974–83, most of the decline in Belgian employment has been concentrated in manufacturing, where hours have fallen 11%, giving a total fall in the labour input of 37.3%. Added to an increase in manufacturing value-added of 15%, the apparent increase in hourly productivity is nearly 85%. At the same time there has been a high increase in real-wages, where the difference between Belgium and the EEC is 20 percentage points, for 1971–80, which is consistent with "a moderate wage elasticity of employment of − 0.5" (Sneessens and Drèze, 1986, p. 97), the authors referring to slope of the labour demand schedule. Thus Belgian manufacturing industry was choked by a combination of labour costs growing at a faster rate than that of competitors *and* a slackening of effective demand, leading to a loss of profits, bankruptcy for the less profitable firms, and loss of markets to foreign competitors. This stylized description of a Keynesian regime can be checked against the data, since in these models time-series estimates for employment (n), labour supply (n^s), 'Keynesian labour demand' (n^d), and 'potential employment' (n^f) are generated endogenously, indicating which of the three standard regimes accounts for Belgian unemployment at any point in the sample, as indicated by Equation 2.19.

The authors pay little attention to the institutional background of the labour market, and its possible impact on explaining why real wages are too high. Given that Belgium has one of the highest union membership rates in the OECD, this might be considered an omission, and of course the standard objection to disequilibrium macroeconomics *per se* is that the modelling of observed wage-price configurations is *ad hoc*. However, alternative possible explanations of short-run disequilibrium are considered, namely that fixed-coefficient technology may act as a short-run constraint on firms expanding output and employment. As we discuss later, high union density is not necessarily a constraint on macroeconomic performance; rather it is a lack of so-called corporatism that correlates with high levels of real wages across countries. Belgium is a case in point, given that national union federations are split along linguistic and religious lines, leading to a fairly decentralized set-up.

The basis of the model estimated is Equations 2.18 explaining the components of aggregate demand, labour demand, wages and prices. Each of the three regimes is specified separately, which if appropriate for *all* the sample period, would form a coherent and simple-to-estimate structure. However, switching-models select regimes endogenously, the role provided by the 'min-condition' given by Equation 2.22, which in this particular application is generalized by a condition that "smooths the kink" (see Chapter 2, p. 34). Assuming that $(N^s/N^d)_i$ has a log-normal distribution over micromarkets, the following employment function (Equation 2.19′) is generated:

$$N = [(N^k)^{-\rho} + (N^s)^{-\rho} + (N^d)^{-\rho}]^{-1/\rho} \qquad \rho > 0.$$

Here N^k refers to all employment determined in micromarkets characterized by a Keynesian regime; similarly N^f for the classical regime and N^s for the Repressed Inflation regime. Associated with the above equation is what Sneessens and Drèze call "the structural unemployment rate at equilibrium", namely the unemployment rate that occurs when $N^s = N^k = N^f$,

$$u^* \equiv (N^s - N)/N^s = 1 - 3^{-1/\rho}$$

The important parameter in this context is ρ, which is "linked to the correlations between the values across micro-markets of the demand for goods, the availability of production capacities and the supply of labour" (Sneessens and Drèze, 1986, p. 104). At one extreme, as ρ tends to infinity, this correlation is almost perfect, or equivalently all the micromarkets are effectively the same and can be represented by one aggregate market. Clearly u^* is zero, and it turns out that the above equation reduces to the more familiar 'min' condition, Equation 2.19. Departures of unemployment from u^* above occur when micromarkets are mismatched, these structural imbalances worsening as ρ tends to zero at the other extreme.[13] However, free estimation of ρ is difficult, although some investigators have used business survey data in this context. Sneessens and Drèze estimate $1/\rho_t$ as a linear function of a constant and trend, indicating that Belgian structural unemployment at equilibrium has risen from 1.39 percentage points in 1955 to 4.5 percentage points in 1982.

Turning now to the equations that determine wages and employment, first n and n^s are observed, in contrast to n^k and n^f. ($n^i = \log N^i$). Aggregate employment, n, enters the model via two so-called technical relationships, which play the same role as the traditional factor demand equations,

$$y - n = A(w - r, t)$$
$$y - k = B(w - r, t),$$

(notation as in Chapter 2, and r is the logarithm of the user-cost of capital). In the short-run labour and capital are assumed to be in fixed proportions, and so A and B are interpreted as fixed constants, which then determine the maximum production possible. The classical regime is defined when this short-run constraint is binding. In the longer run, where capital and labour are substitutable, the two equations have a more familiar form:

$$n - y = -(1 - \alpha)(w - r) + \text{constant}$$
$$y - k = -\alpha(w - r) + \text{constant},$$

being derived from a Cobb–Douglas production function under constant returns to scale.

The wage equation specification is also familiar, and should be regime dependent, but like the price equation, is not:

$$[1 - (1 - \delta)L]w_t = [\alpha - (\alpha - \delta)L]p_t - \beta u_t^r + [\gamma + (\delta - \gamma)L](y - n)_t.$$

13 The extreme values of ρ may be examined with reference to $1/(1+\rho)$, which by analogy is the elasticity of substitution when Equation 2.19′ is interpreted as a CES production function. As ρ tends to ∞, $1/(1+\rho)$ tends to zero; for the production function this is represented by right-angled isoquants (fixed coefficient technology), which by analogy corresponds to the 'kinked' min-condition. In a similar model estimated for the UK by Andrews and Nickell (1986), the underlying assumption is that $(N^s - N^d)_i$ has a normal distribution, with variance σ_x, say. This parameter plays the same role as ρ, since when $\sigma_x = 0$ the employment function reduces to the 'min' condition again, and otherwise it parameterizes the position of the smooth envelope inside 'the kink'. (See also Chapter 3.)

This is an extremely common form of autoregressive/distributed lag relationship between the logarithms of wages, prices, productivity, and excess supply of labour; and corresponds closely to Equations 2.9, 2.12, 2.15 or 2.10. Moreover, if $\beta = 0$, it looks like a rearranged marginal productivity for labour condition.[14]

Thus, in spite of the explicit multimarket framework, the model estimated is quite familiar, except for the novel aggregation of employment over micromarkets. In assessing whether explicit disequilibrium modelling actually provides an adequate description of the data, we note the views of Sneessens and Drèze, who embarked... "with the modest aims of (1) seeking an explanation for differential rate of growth of Belgian real wages in the 1970s; (2) seeking an explanation for the dramatic decline in employment and increase in gross apparent labour productivity of the Belgian manufacturing sector over the past decade; and (3) summarizing what we had learned from the estimation of quantity rationing models" (Sneessens and Drèze, 1986, p. 117). They conclude that: "it is difficult to separate out the respective influences of factor prices (real wages) and effective demand in accounting for the inadequate performance of the Belgian employment since 1974. The only safe conclusion is that both aspects matter" (Sneessens and Drèze, 1986, p. 117). Thus they maintain that the concepts of "potential employment" and "Keynesian labour demand" provide a convincing, if not incomplete story of Belgian macroeconomic performance. Our view, bearing in mind the estimation of a similar kind of model for the UK (Andrews and Nickell, 1986), is that the econometric difficulties noted in Chapter 2 militate against any real value-added.

Finally, in completing our selective survey of labour market models applied to a single country, note that Driehuis (1986) also presents a multimarket disequilibrium model for Holland. Turning to applications of model (F), as formulated by Layard and Nickell (1986) for the UK. In the recent *Economica* Special Issue Modigliani *et al.* (1986) present such an application for Italy, as do Dolado *et al.* (1986) for Spain.

7.4 MULTI-COUNTRY STUDIES: SOME EXAMPLES

Corresponding to the last column of our schematic matrix, the last two sections consider a selection of econometric studies that have been applied to more than

[14] In particular, notice that first if $\alpha = 1$ all price effects are passed fully on to wages, and so the equation effectively determines the *level* of real wages in terms of unemployment and the *level* of productivity. In the long run the equation says that labour's share is determined by unemployment with an elasticity of $-\beta/\gamma$. If also $\delta = 0$, then the model determines the change in real wages, as in Equation 2.15 which is then determined by labour market tightness (with an elasticity of $-\beta$) and changes in productivity (elasticity γ). In the long run, with $\delta = 0$, it simply defines the natural rate of unemployment. In a sense the equation is a structural form proxy for a reduced form wage equation, in so far as unemployment is capturing supply-side effects, and productivity demand-side effects. Thus if unemployment is insignificant, $\beta = 0$ (and also $\alpha = 1$) then the equation looks a dynamic form of the marginal productivity condition for a CES production function, under constant returns (Equation 2.4).

one country. Referring back to our discussion of Section 7.2, in this section we consider studies where countries simply represent "one among several essentially identical regressions", and in the next, studies where institutional differences are considered explicitly. Three sets of hypotheses appear to be of some international interest, namely: whether real wages and employment are independent; whether real and/or nominal wages exhibit rigidity; and whether there is a role for aggregate demand in inducing shifts in the labour demand schedule. Of course these hypotheses are not unrelated. However, of the three, only the first is usually couched in terms of the *joint* determination of wages and employment.

7.4.1 The independence of real wages and employment hypothesis

As discussed in Chapter 2, this hypothesis is usually tested in the framework represented by model (G), and is particularly favoured in North America. In this context the findings of Neftci (1978) and Sargent (1978), namely that employment and the real wage are negatively associated, with the causality running from real wages to employment, were well established at the turn of the decade.[15] Since most of the accumulated evidence was for the US and Canadian economies only, Geary and Kennan (1982) extend the investigation using quarterly data for the manufacturing sectors of 12 OECD countries.

The Nefti and Sargent evidence is usually interpreted as the identification of the labour demand schedule. Geary and Kennan are concerned as to whether this is a correct interpretation, and argue that, like other investigators (see Section 7.2): "If [employment] and [wages] in fact satisfy some economic relationship, such as a dynamic demand curve, one would expect this relationship to hold across countries. This suggests that the data should be pooled in case the relationship is not strong enough to appear significant in the sample available for any one country" (Geary and Kennan, 1982, p. 857). Thus they report a statistic that is an average, over the 12 countries, of the one of two statistics they compute for each country.[16,17]

Geary and Kennan find "the significant employment-real wage relationship found by Neftci for the United States is not a regular feature of business cycles in most economies", also failing to find Neftci's and Sargent's result in US data.

[15] Neftci regresses both employment and unemployment on a distributed lag of the real wage, using monthly US data, and correcting for serial correlation. He finds a long-run relationship between employment and the real wage, which gets weaker with fewer lags of the regressand, in fact being positive in the static case.

[16] Geary and Kennan compute two sets of asymptotically equivalent statistics. The first is related to Sim's (1972) test for unidirectional causality, and essentially tests for $a_{12} = 0$ in the top row of Equation 2.21 and $a_{21} = 0$ in the bottom row, where the data for employment and the real wage have been detrended and deseasonalized. The second method involves computing innovations of these data from an estimated AR (10) process, and computing a summary statistic of the cross-correlogram (Haugh, 1976).

[17] This pooling technique is extended by other investigators whereby parameter estimates, and their *t*-ratios, for a common specification applied to a number of countries, are averaged. This dubious practice leads to concepts such as an OECD Phillips curve (Grubb, 1986).

In explaining (encompassing) this in the context of their own evidence, they note the use of different sample periods, and the use of different deflators in the real wage measures. When Neftci deflates using the consumer price index a negative, but unstable, relationship is found, whereas Geary and Kennan themselves use the "more appropriate" wholesale price index and find no relationship. These general conclusions are supported by the pooled statistics across countries. Finally, Geary and Kennan find that Belgium and Canada provide the best evidence in support of the independence hypothesis, and Ireland and NZ the worst, but they provide no explanations for these results, nor mention any possible differences in labour market structure, since under the null-hypothesis of independence all differences in the results across countries can only arise from random sampling error.

Geary and Kennan's results are of some interest since they highlight a common problem encountered when estimating vector autogressions. This may be illustrated by considering a bivariate plot of the real wage and employment, which typically reveals some periods of distinct positive association, some negative, and some periods neither, giving a rather hazy overall picture (see Andrews, 1987b, Fig. 1, for the UK). This explains the sensitivity of results to sample and country, and arises because labour demand shocks have dominated supply shocks for some periods, but not others, depending on whether they have foreign or domestic source, on whether the macropolicy mix is 'fiscal' or 'monetary', and so forth. A simple econometrics textbook discussion of identification suggests a 'hazy' plot is necessary if *both* the downward-sloping demand curve and the other upward-sloping schedule are to be identified, whereas a fairly systematic covariation indicates a lack of identification. Moreover, when some distinct association is found in the data, often it is not clear the slope of which schedule is actually being estimated, as the lag structure of the autogression could be proxying omitted variables from either equation. Thus one view is that this particular methodology is at best a useful descriptive device, saying little about the actions of labour market agents, lending itself more easily to examining issues such as wage rigidity, to which we now turn.

As discussed briefly in Chapter 2 wage rigidity, be it nominal or real, can be investigated in one of at least three ways, namely: by estimating the autoregressive structure of wages; or more simply by calculating sample standard deviations of the rate of change of wages; or by estimating a wage equation. Typically the third of these is used, and the most common specification is the simple expectations-augmented Phillips curve, Equation 2.9. However, there is some confusion as to precisely how wage rigidity is defined, although the size of the effect of unemployment on the rate of change of nominal wages appears to be important, as is the magnitude of the coefficient on (expected) inflation. Since in this chapter we are only examining models that jointly determine both wages *and* (un)employment, we shall not investigate this matter further, except to note that further confusion arises as to whether real or nominal wage rigidity are mutually independent. For example "some authors (e.g. Sachs, 1979) have defined real wage rigidity as the opposite of nominal wage rigidity, with the US nominally rigid and Europe really rigid" (Grubb *et al.*, 1983, p. 11), whereas Branson and Rotemberg (1980) manage to derive measures of both by estimating *one* wage equation, for a number of countries. Grubb *et al.*'s point has some merit, since it is difficult to see how two separate hypotheses about two different endogenous variables, w and $w - p$, can be tested by estimating one equation, which of course can only determine one of them. However, what *is* certain is that estimates of real/nominal wage rigidity are important in making inter-country comparisons on labour market performance. (See below; and also Sachs, 1983; Taylor, 1983; Newell and Symons, 1986; and Grubb, 1986.)

We now consider *structural* models that jointly determine wages, employment, and/or unemployment, as applied in multi-country studies. Following our discussion in Section 7.2, our theme in the rest of this chapter is to examine the degree to which institutional differences across countries are explicitly accounted for, or even just considered, in the formal analysis.

Newell and Symons (1987) take a sizeable step in the right direction. They estimate a two-equation model applied to data for 16 OECD countries, corresponding to model (B) of Chapter 2, although in the first instance the wage and employment equations estimated are the more general Equations 2.20 and 2.8 respectively. The restrictions corresponding to perfect competition are tested. In particular none of the following variables have a consistent effect on the real product wage, *across countries*, namely: industrial skill mismatch, union power, industrial conflicts, unanticipated inflation, unemployment benefits, the price of raw materials, and labour market taxes. The incidence of taxation is completely on the worker, which implies an inelastic labour supply schedule. However, they discover some short-run real wage rigidity insofar as wages have been pushed above their long-run market-clearing level by changes in taxation, relative prices and what the authors call 'explosion'. In examining the aggregate demand hypothesis, they conclude: "We believe this work, which has tested every combination of these demand variables we could think of, has re-established the results reported in Symons (1985b) and Symons and Layard (1984). In individual countries it is possible to obtain at times significant results, but we have found no uniformity in the international data which suggests variables additional to the wage as proximate determinants of employment" (Newell and Symons, 1987, p. 15). See also the contrasting evidence of Bean *et al.* (1986) and Bruno (1986). The issue here is whether one would really expect the range of variables found to be important for a UK wage equation (for example, Nickell and Andrews, 1983) to be significant for each and every OECD country.

Having applied the same basic market-clearing structure throughout, the authors then characterize the aggregate labour markets of each economy by the estimated values of four basic parameters, these being the slopes of demand and supply/wage, the employment lag in the demand equation, and the wage lag in the 'other' equation. These determine how quickly a particular economy will respond to wage and employment shocks. It turns out that most of the variation across countries is attributable to the slope of wage/supply equation and the employment lag. Each country is then classified in these two dimensions, the authors suggesting that the economies with the least unpleasant unemployment experience in the 1970s were those whose wage/supply schedule is steep (Japan, Switzerland, Sweden, Italy, Australia), since unemployment is able to exert a downward pressure on wages. Of the rest unfortunately most also exhibit strong persistence in the employment equations (Canada and 'Europe'); the two exceptions being Austria and US, the latter being blessed with particularly fast employment adjustment. However, the authors do not attempt to correlate these inter-country differences picked-up in the regression analysis with differences in so-called background. For example, the US case is not unsurprising, since it is well known that labour is generally less costly to hire and fire, consistent with the observation that in terms of labour market characteristics, the US is not a typical OECD country.

7.5 INSTITUTIONAL CHARACTERISTICS

The view noted earlier, that when modelling and comparing the labour markets of more than one country, inter-country differences in empirical results should be accounted for by considering the possibility that institutions, custom, and practice (what we have called 'background') may also differ from country to country, has been generally ignored in the studies considered thus far. This section examines those that are the exception. Up to now we have been somewhat vague as to what 'background' really is; we are now more explicit, the following list identifying some of the more important characteristics:

(i) Macroeconomic objectives
(ii) Whether bargaining takes place at the national, local or company level;
(iii) Whether the power of labour organizations ('unions') is at the local or national level, which is related to the power of local union stewards, for example;
(iv) The extent to which unions are involved in wage determination;
(v) The extent to which there is co-ordination with employers, by the extent to which employers' organizations have a role in policy-making; patterns of industrial conflict as reflected in enforcement strategies
(vi) Whether unions and employers have strong affiliations to political parties;
(vii) Other objectives of unions, both individually, and as a national organization;
(viii) Whether incomes policies are a part of general macro-policy, or whether there are any other institutional reasons why wages are indexed to prices;
(ix) The existence of wage tribunals or councils (e.g. 'The Commission' in Australia);
(x) The financing arrangements of firms;

(xi) Whether there exist particular incentive schemes for workers, for example profit-sharing schemes, or bonus systems (as in Japan);
(xii) The characteristics of the contracts made between workers and firms;
(xiii) The incidence of non-wage labour costs, their relationship to the tax-benefit system, and other costs to labour adjustment;
(xiv) General work-place attitudes, for example towards sex-discrimination, child-rearing, and so forth.

Most are self-explanatory, although some comments are called for. For example it is possible that different countries have different macroeconomic objectives independently of whether unions are able to, or perceive, a role in policy-making. On (iv), it should be noted that the extent to which unions are involved in wage determination, or 'coverage', is not the same as membership or 'density'. The role of unions clearly depends on where bargaining takes place, (ii) above, where they are able to exert influence, (iii), and the labour market structure (e.g. competitive or oligopolistic). If it is true that the larger unions have a greater potential for wielding power, their actual influence depends on the identity of the governing parties (vi), which is related to their objectives (vii). For example unions may also be concerned with the pay and conditions that their rivals enjoy, suggesting that there is more to modelling the behaviour of trade unions than by assuming a bargaining set-up where unions have utility over only real wages and employment. Gylfason and Lindbeck (1984) formalize a rigid industrial wage structure by applying a Cournot-style analysis to an oligopolistic union structure.

As already hinted at, the way in which the labour input (man-hours) responds to changes in demand is a key determinant of macroeconomic performance, and is determined by many factors which vary from country to country. Central to these are non-wage labour costs, the whole area admirably surveyed by Hart (1984). The labour input can be altered in one of two ways: via hours of work, leading to either short-time working or overtime working; or via heads, leading to temporary layoffs. Clearly the relative costs associated with these alternatives determines their incidence. Moreover, non-wage labour costs are bound up with the tax/benefit system, since the firm in part provides for the unemployed worker via unemployment insurance, a classic example being the role of the Italian Wage Supplement Fund in subsidizing temporary layoffs. (Incidentally this establishes a second channel, on the *demand* side, whereby unemployment is influenced by benefits, in addition to the usual supply-side effect.) The role of unions in this relationship has also been identified, their presence shifting the incidence in favour of adjustment via heads as opposed to hours, as has their role in firm level (employment) flexibility, where "factors like job security rules, craft demarcations and work patterns help determine the level of employment associated with any particular stance of macroeconomic policy" (Metcalf, 1986, p. 26).

Turning now to the contracts made between workers and employers, (xii) above, their characteristics also vary from country to country. In particular these are whether they are written, or informal; their duration; whether they are open-ended; and whether their timing is synchronized across the whole economy. Clearly differences in job tenure across countries may be attributable to this, as well as (xi), (xiii) and (xiv).

Assessing the impact of each of the above fourteen characteristics on macroeconomic outcomes is difficult within any one country, although slightly less so across countries, essentially for data reasons. This is because where a particular characteristic can be quantified – of course this is not always possible, as in (v) above – the data cannot be collected for a complete time-series that would then be used in the type of regression exercises discussed in the previous section. This proposition is illustrated in Hart (1984, chs 2 and 3), where in spite of the detailed information on the range of non-wage labour costs for four OECD countries (UK, US, West Germany and Japan), complete time-series are not available. Two alternative approaches may be identified. The first is to predict the impact of a characteristic on observable aggregates, and then assess whether variation across countries correlates in the 'right' direction. This typically requires examining the time-series behaviour of wages, hours, employment and so forth, which implicitly assumes that these institutional features are constant over time, when in fact often they are not.

An example is provided by Gordon's analysis of *nominal* wage rigidity, something that is not easily explained by economists. The motivation behind the exercise is similar to the view expressed by Solow earlier in this chapter:

> *"American economists, whose theoretical ingenuity is matched by their institutional chauvinism, have not succeeded in developing an adequate economic explanation of labour-market arrangements; their theories that purport to explain wage stickiness are mainly based on elements that do not differ across nations and thus have little potential for explaining why the degree of wage flexibility is much greater in some places than in others."*
>
> (Gordon, 1982, p. 14)

Thus Gordon's view is that a detailed study of the institutional background is absolutely necessary in explaining the phenomenon (see also p. 171). The *real* problem concerns nominal rigidity itself, since there is a lack of unanimity in classifying those economies which *are* nominally rigid, and those which are not, because there is confusion as to how the concept is established empirically. Gordon's own evidence adds to this state of affairs, partly challenging the received wisdom of Bruno and Sachs, and Branson and Rotemberg, namely that nominal wages in the US are rigid whereas European economies exhibit real wage rigidity. Gordon compares the volatility of nominal wages (relative to other labour-market aggregates) between the US, Japan, and the UK, using post-war quarterly manufacturing data. By calculating the standard deviation of the change of nominal wages, and by regressing labour-market aggregates on the changes in nominal GNP, he reports that nominal wages fluctuate more than employment and hours for UK and Japan, and vice versa for the US; that employment fluctuations are roughly the same for Japan and the UK; and that *real* wages *also* fluctuate more in the UK and Japan than in the US. It is this third finding that is controversial, (see also Coe and Gagliardi, 1985), and of course questions the sense in which real and nominal wage rigidity are the opposite sides of the same coin.

On the basis of these 'facts', Gordon considers what economic theory has to say about the *observed* organization of the labour markets of the three countries, providing a detailed discussion of their background, focusing on job tenure, seniority rules, and contract durations. It is argued that the introduction of the three-year staggered contract in the US in 1948 helps explain US nominal rigidity; that the historical and cultural background of Japan has contributed towards avoidance of conflict wherever possible, explaining shorter contract durations in Japan and a high degree of nominal wage responsiveness. The UK is the odd man out, with well-publicized industrial strife but also short contract durations. On this hypothesis the puzzle is why the UK did not see the introduction of written and long-term contracts at some point in history, a question more for economic historians than macro-theorists. In an OECD context, however, Italy is the only European country with long contracts, re-establishing the US as an outlier, although none of these three economies are typical of the OECD. In spite of its commendable descriptive approach, when the same methodology is subsequently used by others the same problem in interpreting the observed macro-aggregates recurs. For example, Metcalf (1986, p. 4), in his up-to-date analysis of labour market flexibility, reveals some lack of confidence in Coe and Gagliardi's (1985) estimates of the non-accelerating inflation rates of unemployment (NAIRUs) for nine OECD countries.

A second approach in dealing with the lack of time-series data on so-called background is to actually construct (time invariant) measures thereof for each country. The investigator then either (i) examines the correlations (total or partial) between these measures and observed aggregates across countries; or (ii) generates other summary measures from the standard multi-country time-series regression-based analysis discussed in the previous section, a good example being Newell and Symon's four key parameters, which are then correlated with the aforementioned constructs of background.

Bruno and Sachs (1985, ch. 11) first generated and made use of such data, as a by-product of a study whose primary objective was to explain the different degree of stagflation experienced by the OECD countries in response to the supply shocks of the 1970s. Their basic hypothesis is that the real wage is central to successful macro-adjustment, and that certain institutional arrangements appear to be conducive to real wage moderation after a (supply) shock. These are: centralized wage negotiations, i.e. (ii) above; high degree of co-ordination with employers, (v); low shop-floor presence of unions, (iii); low wage indexation (viii); infrequent collective bargaining, and, if decentralized, bargaining should be unsynchronized (xii). Thus they identify two distinct measures for summarizing background, namely 'corporatism' and the degree of indexation. The former "is defined as a mode of social organisation in which functional groups rather than discrete individuals wield power and transact affairs" (Bruno and Sachs, 1985, p. 222), and essentially is developed as a manifestation of consensus by earlier political scientists. An index (a cardinal measure) and a set of rankings (ordinal) of corporatism for each of 17 OECD countries are then derived. The high rankings for Austria (rank = 1), West Germany (2), Holland (3), Norway and Sweden (both $4\frac{1}{2}$) are not surprising, nor are the rankings for US (17) and Canada (16). Japan and the rest of Europe are in the middle. Bruno and Sachs then report partial correlations between the index and inflation minus growth (the so-called 'misery' index) and between the index and the wage gap. In offering some explanations for this latter negative association, they suggest that labour will concede some of its share of the cake, in wages, for a greater stake in policy making, reflecting a conflict in objectives as noted above. More important is the idea that the corporatist structure is better able at absorbing shocks, through consensus, thereby ameliorating the potential delitirious effects of inter-union rivalries. However, this association *does* rest on excluding the US from the sample, which being non-corporatist exhibits low wage-gaps and also a low indexation of wages to prices, suggesting that corporatism is only a necessary condition for superior macroeconomic performance. Bruno and Sachs also construct indices of nominal wage responsiveness, based on the length of wage-agreements, the degree of explicit indexation, and the extent to which wage setting is highly synchronized. This measure correlates with the two other measures noted earlier, the coefficient on inflation in the APC and the more simple-minded standard deviation of nominal wage changes. Having classified the 17 countries in these two dimensions, namely corporatism and nominal wage responsiveness, they are then correlated in a similar manner to Newell and Symons, finding no particular relationship between the two.

Bean *et al.* (1986, section 4) also make use of the corporatism ranking data, using the Layard–Nickell model of Chapter 6 of this book. (See also model (F) of Chapter 2.) In estimating wage and employment equations for nineteen OECD countries, it turns out there is a downward-sloping labour demand schedule in all except the US. (A respecification does unearth the required negative coefficient, and so the authors conjecture an explanation based on the US being the most 'closed' of the OECD countries, the evidence of Nickell and Symons (1986) notwithstanding.) Aggregate demand has a significant positive effect except for Australia, Japan and Sweden; and the wage equation has the correct slope for all countries except Italy and the US.

Having successfully applied the same model to each OECD country, the authors seek to relate variation in estimates across countries to the Bruno and Sachs corporatism rankings. Because it is an ordinal measure, and because it only varies across countries, not time (a so-called fixed effect), it cannot be introduced as a regressor-type variable. Bean *et al.* extend Newell and Symons' work by associating (non-parametrically) corporatism with some important estimated parameters of each country's labour-market model, namely the short- and long-run slopes of the wage equation, the effect of supply shocks in the wage equation, the speed of adjustment in the wage and employment equations, and observed unemployment. The general findings are that "the evidence seems to support the notion that structural differences in labour markets can be related to national differences in institutional and social characteristics", and "the labour market as a whole seems to adjust more quickly in economies that are more corporatist in nature", noting that also the results are not particularly supportive of the idea that unions *per se* inhibit the efficient working of the labour market.

We conclude this section by suggesting that Bean *et al.* go some considerable way to answering Solow's critique noted in Section 7.2. However, there are some fairly obvious problems, namely that measuring the corporatism-consensus nexus is subject to considerable value-judgement, and the more substantive point that corporatism is not time-invariant. As Metcalf (1986) notes "the definition of corporatist and non-corporatist episodes is a matter of fine judgement. Further there may be lags in the impact of corporatism on labour market variables" (Metcalf, 1986, fn 1). Nonetheless this type of analysis is an important addition to the more descriptive analysis of the Gordon and Metcalf genre.

7.6 CONCLUDING COMMENTS

The objectives of this exercise were carefully set out in Section 7.1, and do not bear repetition. To conclude this chapter we briefly set down our impressions of, on the basis of a partial survey of the literature, how wages and employment are modelled from country to country. Three points stand out.

First, in considering the relative frequencies of the cells of our schematic matrix, some model-types appear to be prevalent in certain groups of countries than they do in others. Although our evidence is extremely impressionistic, it does appear that although model (A) is still extremely common, it is waning in more academic studies. It is interesting to examine the small sample of single-country studies, with native authorship, reported in the *Economica* Special Issue on OECD unemployment. Of the nine, four are of model (A) type, namely Australia, Austria, France and Japan. However, more structured models are becoming popular, as illustrated by the applications of the Layard–Nickell approach to the UK, Italy, and Spain. Also the European preference for the more complicated disequilibrium approach is reflected in the applications to Holland and Belgium. Clearly this use of disequilibrium is consistent with the view that Europe exhibits real wage rigidity in contrast to the nominal rigidity of the US, also explaining the popularity of the non-homogenous APC in North America, i.e. models (A) and (D), particularly as Lucas and Rapping's (1969) simple but elegant model also incorporates such a feature. Market-clearing, i.e. model (B), makes much more sense if it is nominal, not real, wages that are rigid, as a rapid burst of inflation can remove the wage-gap and associated unemployed labour if required.

Second, recent work on cross-country explanations of macroeconomic performance have started to move away from the blanket application of the same model-type to all the countries *without* consideration as to how different parameter estimates might be explained by, or be associated with, different 'background'. Although there are considerable data problems in establishing these links in the usual regression-based analysis, at least the characteristics derived from a typical econometric study of one or more countries are now being set against the range of institutional features noted in the previous section. A comprehensive model-based check-list relevant for inter-country comparisons might be as follows:

(i) The slopes of the long-run employment/labour demand schedule, as represented by Equation 2.1;
(ii) The dynamic adjustment of employment in Equation 2.1;
(iii) Whether Equation 2.1 sustains aggregate demand measures.

Subsidiary features not considered in this chapter relate to whether men and hours are perfect substitution, the elasticity of substitution between labour and other factors, whether manufacturing behaves differently from the whole economy, and so forth.

(iv) Whether the 'other' schedule is best represented by Equations 2.9, 2.12, 2.13, 2.15, 2.15^1, or 2.20. This involves issues such as;
(v) Short-run homogeneity;
(vi) The most appropriate way of modelling labour-market slackness;
(vii) The slope of the schedule over the short- and long-runs;
(viii) The speed of wage adjustment.

Finally, issues that arise from taking both schedules together, as discussed in the main body of this chapter, are:

(ix) Whether wages and employment are determined independently of each other;
(x) Whether positive wage-gaps are observed;
(xi) Whether the market as a whole is able to respond quickly to shocks.

This again addresses whether the wage exhibits either nominal or real rigidity, but in a simultaneous equations context.

Our final comment relates to the degree to which econometric practice varies from country to country, which also influences the way regression evidence is weighted and interpreted. What is clear is that European, North American, and UK practice differ. The emphasis on non-structural methods is greater in North America. Moreover, the econometric practice of Bruno and Sachs might be considered less rigorous than is typically found in the UK, particularly relating to certain aspects of specification-search, the range of variables chosen, lags included, and the general use of diagnostic checks. Of course 'rigour' to one investigator is tantamount to 'data-mining' for another, which underpins the difference between the two schools of thought identified earlier relating to multi-country studies (Section 7.2). This distinction is less important in the

European disequilibrium approach, where because of the intrinsic complexity of the models, successful implementation is more difficult than in the standard LDSEM approach (Section 2.1), and so diagnostic practice is an added luxury. As always there is a tradeoff between parsimony and reality.

ACKNOWLEDGEMENTS

Help from Hali Edison, Alan Harrison, Wiji Narendranathan, and especially Ben Knight, with directing me to some of the literature that I would otherwise have missed is gratefully acknowledged, as are the comments of Alan Harrison, Ben Knight and Mark Stewart on earlier drafts. Mike Beenstock's help in creating the structure of Chapters 2 and 7 is also acknowledged.

REFERENCES

Altonji, J. G. (1982) The intertemporal substitution model of labour market fluctuations: an empirical analysis, *Review of Economic Studies*, **49** (supplement), 783–824.

Altonji, J. G. and Ashenfelter, O. (1980) Wage movements and the labour market equilibrium hypothesis, *Economica*, **47**, 217–45.

Andrews, M. J. (1987b) The aggregate labour market – an empirical investigation into market clearing for the UK, *Economic Journal*, **97**, 157–76.

Andrews, M. J. and Nickell, S. J. (1986) A disaggregated disequilibrium model of the labour market, *Oxford Economic Papers*, **38**, 386–402.

Ashenfelter, O. and Card, D. (1982) Time-series representations of economic variables and alternative models of the labour market, *Review of Economic Studies*, **49** (supplement), 761–82.

Ashenfelter, O. and Card, D. (1986) Why have unemployment rates in Canada and the United States diverged?, *Economica*, **53** (supplement), 171–96.

Bean, C. R., Layard, P. R. G. and Nickell, S. J. (1986) The rise in unemployment: a multi-country study, *Economica*, **53** (supplement), 1–22.

Board of Governors of the Federal Reserve System (1984) *The US Economy In An Interdependent World*, Washington.

Branson, W. H. and Rotemberg, J. J. (1980) International adjustment with wage rigidity, *European Economic Review*, **13**, 309–32.

Brooks, S. H. (1984) Macroeconometric models: theory and practice, in *Papers in Applied Business Economics*, National Association of Business Economists, November.

Bruno, M. and Sachs, J. D. (1985) *Economics of Worldwide Stagflation*, Blackwell, Oxford.

Bucher, A. and Rossi, V. (1985) Analysis of stabilisation mechanisms of macroeconomics models: a comparison of the Eurolink models, EEC (mimeo), February, Brussels.

Caves, R. E. and Krause, L. B. (eds) (1984) *The Australian Economy: A View from the North*, Brookings Survey of the Australian Economy, George Allen and Unwin, Sydney.

Coe, D. and Gagliardi, F. (1985) Nominal wage determination in ten OECD economies, OECD, Economics and Statistics Department, working paper 19, March, Paris.

Dolado, J. J., Malo de Molina, J. L. and Zabalza, A. (1986) Spanish industrial unemployment: some explanatory factors, *Economica*, **53** (supplement), 313–34.

Driehuis, W. (1986) Unemployment in the Netherlands: 1960–83, *Economica*, **53** (supplement), 297–312.

Eliasson, G. (1986) *The Firm and Financial Markets in the Swedish Micro-to-Macro Model – Theory, Model, and Verification*, The Swedish Institute for Economic and Social Research, Stockholm.
Franz, W. and Konig, K. (1986) The nature and causes of unemployment in the Federal Republic of Germany since the 1970s: an empirical investigation, *Economica*, **53** (supplement), 219–44.
Geary, P. T. and Kennan, J. (1982) The employment–real wage relationship: an international study, *Journal of Political Economy*, **90**, 854–71.
Gordon, R. J. (1982) Why US wage and employment behaviour differs from that in Britain and Japan, *Economic Journal*, **92**, 13–44.
Grady Economics and Associates Ltd (1985) The state of the art in Canadian macro-economic modelling (mimeo), March, Canada.
Gregory, R. G. (1986) Wages policy and unemployment in Australia, *Economica*, **53** (supplement), 53–74.
Grubb, D. (1986) Topics in the OECD Phillips Curve, *Economic Journal*, **96**, 55–79.
Grubb, D. Jackman, R. and Layard, P. R. G. (1983) Wage rigidity and unemployment in OECD countries, *European Economic Review*, **21**, 11–39.
Gylfason, T. and Lindbeck, A. (1984) Union rivalry and wages: an oligopolistic approach, *Economica*, **51**, 129–40.
Hamada, K. and Kurosaka, Y. (1986) Trends in unemployment, wages and productivity: the case of Japan, *Economica*, **53** (supplement), 275–96.
Hart, R. A. (1984) *The Economics of Non-wage Labour Costs*, George Allen and Unwin, London.
Haugh, L. D. (1976) Checking the independence of two covariance-stationary time series: a univariate residual cross-correlation approach, *Journal of the American Statistical Association*, **71**, 378–85.
Helliwell, J., Sturm, P., Jarrett, P. and Salou, G. (1985) Aggregate supply in Interlink: model specification and empirical results, OECD, Economics and Statistics Department, working paper 26, November.
Kniesner, T. J. and Goldsmith, A.H. (1986) Does the labour market clear? A survey of the evidence for the United States, *Research in Labour Economics*, **7**, 209–56.
Layard, P. R. G. and Nickell, S. J. (1986) Unemployment in Britain, *Economica*, **53** (supplement), 121–70.
Lucas, R. E. and Rapping, L. A. (1969) Real wages, employment and inflation, *Journal of Political Economy*, **77**, 721–54.
Malinvaud, E. (1986) The rise of unemployment in France, *Economica*, **53** (supplement), 197–218.
Metcalf, D. (1986) Labour market flexibility and jobs: a survey of evidence from OECD countries with special references to Great Britain and Europe, Centre for Labour Economics, London School of Economics, discussion paper 254, October.
Mitchell, D. (1984) The Australian Labor Market in *The Australian Economy: A View from the North* (eds R. E. Caves and L. B. Krause), Brookings Survey of the Australian Economy, George Allen and Unwin, Sydney.
Modigliani, F., Padoa Schioppa, F. and Rossi, N. (1986) Aggregate unemployment in Italy: 1960–83, *Economica*, **53** (supplement), 245–74.
Neftci, S. N. (1978) A time-series analysis of the real wages-employment relationship, *Journal of Political Economy*, **86**, 281–91.
Newell, A. and Symons, J. S. V. (1986) The Phillips Curve is a real wage equation, Centre for Labour Economics, London School of Economics, discussion paper 246, July.
Newell, A. and Symons, J. S. V. (1988) Wages and employment in the OECD countries, *Economic Journal* (forthcoming).
Nickell, S. J. and Symons, J. S. V. (1986) The real wage-employment relationship in the

United States, Centre for Labour Economics, London School of Economics, discussion paper 264, November.

Pichelmann, K. and Wagner, M. (1986) Labour surplus as a signal for real-wage adjustment: Austria 1968–84, *Economica*, **53** (supplement), 75–88.

Rosen, H. S. and Quandt, R. E. (1978) Estimation of a disequilibrium aggregate labour market, *Review of Economics and Statistics*, **60**, 371–9.

Sachs, J. D. (1979) Wages, profits, and macroeconomic adjustment: a comparative study, *Brookings Papers on Economic Activity*, **2**, 269–333.

Sargent, T. J. (1978) Estimation of dynamic labour demand schedules under rational expectations, *Journal of Political Economy*, **86**, 1009–44.

Sims, C. A. (1972) Money, income and causality, *American Economic Review*, **62**, 540–52.

Sneessens, H. (1983) A macroeconomic rationing model of the Belgian economy, *European Economic Review*, **20**, 193–215.

Sneessens, H. R. and Drèze, J. H. (1986) A discussion of Belgian unemployment combining traditional concepts and disequilibrium economics, *Economica*, **53** (supplement), 89–120.

Solow, R. M. (1986) Unemployment: getting the questions right, *Economica*, **53** (supplement), 23–34.

Symons, J. S. V. (1981) The demand for labour in British manufacturing, Centre for Labour Economics, London School of Economics, discussion paper 91.

Symons, J. S. V. (1985a) Employment and the real wage in Australia, in *The Structure and Duration of Unemployment in Australia* (ed. P. A. Volker), monograph series 6, Bureau of Labour Market Research.

Symons, J. S. V. (1985b) Relative prices and the demand for labour in British manufacturing, *Economica*, **52**, 37–49.

Symons, J. S. V. and Layard, P. R. G. (1984) Neoclassical demand for labour functions for six major economies, *Economic Journal*, **94**, 788–99.

Taylor, J. B. (1983) Comment on 'Wage rigidity and unemployment in OECD countries', by D. Grubb *et al. European Economic Review*, **21**, 45–49.

Trivedi, P. K. and Alexander, J. N. (1986) Incorporating international competitiveness into the demand for labour function: some issues of specification and interpretation, Australian National University (mimeo), June.

van Wijnbergen, S. (1985) International repercussions of trade intervention and macroeconomic policies in the industrial countries: a developing countries perspective, *Economic Policy*, **1**, 82–137.

Wallis, K. F. (ed.), Andrews, M. J., Bell, D. N. F., Fisher, P. G. and Whitley, J. D. (1984) *Models of the UK economy: A Review by the ESRC Macroeconomic Modelling Bureau*, Oxford University Press, Oxford.

Wallis, K. F. (ed.), Andrews, M. J., Fisher, P. G., Longbottom, J. A. and Whitley, J. D. (1986) *Models of the UK Economy: A Third Review by the ESRC Macroeconomic Modelling Bureau*, Oxford University Press, Oxford.

Index

Page numbers followed by n refer to footnotes; pages numbered in *italics* contain figures.

Adjustment
  adaptive 3
  equations 34, 86, 87, 88–9
  lags 15
Adverse selection model 11–12
Agency shop arrangements 110
Aggregation 1, 2
Alexander, J. N. 178
Alogoskoufis, G. S. 33
Altonji, J. G. 33, 179, 180, 181
Andrews, M. J. 25–48, 33, 34, 41n, 59, 60, 65, 169–97
Arbitrage condition 153
Arbitration 177
Ashenfelter, O. 179, 180, 181
Australia 176–8
Australian Conciliation and Arbitration Commission 177, 188
Autocorrelation 126
Azariadis, C. 43

Background 188–93
Baily, M. N. 43
Bain, G. 116
Bargaining 2, 13, 14, 189, 191
  models 40–1, 59
  Nash bargain 40
Barro, R. 59, 68
Barro, R. J. 37, 43
Batchelor, R. A. 120
Bean, C. R. *et al.* 169n, 187, 192–3
Beckerman, W. 164
Beenstock, M. 3, 19n, 71–104
Beenstock, M. *et al.* 72, 81
Belgium 181–4, 186
Benassy, J. P. 49
Benefits
  earnings related supplement 114, 116
  retirement 76, 84
  simulated increase 95, *96*
  toughness in denying 126
  see also Unemployment benefits
Benjamin, D. 123
Bils, M. 147n
Blanchard, O. 19, 150, 164
Bliss, Christopher 1–23
Bowden, R. J. 34
Branson, W. H. 187, 190
Brookes, S. 128
Brooks, S. H. 174–5
Bruno, M. 19, 28, 35, 73, 176, 178, 187, 190, 191–2
Bucher, A. 176n
Bureau of Economic Analysis, Commerce Department 174n

Cambridge Growth Project (CGP) 174n
Canada 175, 185, 186
CANDIDE, Economics Council of Canada 175n
Capital stock 27, 28, 55, 75, 82
  simulated increase 93, *94*
Card, D. 179, 181
Carlin, W. 155n
Centre for Labour Economics 169n
Chase Econometric Associates Inc. 174n, 175n
CHASE Economic Model 175n
Child dependency ratio 77, *78*, 85
Chow test 126

Citibank Information Services Inc. 174n
City University Business School *see* CUBS model
Clower, R. W. 49, 52
Cobb-Douglas production function 26–7, 28, 36, 44, 175, 183
Coe, D. 191
Coefficient constancy, F-test for 126
COMET model 176n
Competition
 behaviour 2
 imperfect 29, 39–42, 45, 75, 145–67
  empirical analysis 160–6
  employment behaviour 147–9
  macromodel 152–60
  pricing behaviour 146–7, 148–9
  wage determination 149–52
 perfect 26–8, 74–5
Contracts 12–13, 124–5, 189–90
de Cooman, E. *et al.* 77
Corporatism 191–2
Cournot-Nash industries 147–8, 189
Cowling, K. 147n
CUBS model 81, 88, 174n

Data Resources Inc. 174n
Data-mining 172, 194
Deaton, D. 3
Deflationary policies 140
Deflator
 common price 127
 factor cost (TPI) 83
 gross domestic product (GDP) 82
 market price total final expenditure (TFE) 82–3
Demand for labour 2–3, 26–30, *64*
 conditional schedule 28
 constant elasticity 147
 disequilibrium approach 55–8
 effective 38
 neoclassical model 72–5
  by business sector 81, 82, 83, 84
 new classical model 106, 120–39
  annual data 128–9, 130
  elasticity 125, 126
  empirical work 121–2
  planned output and 121
  speed of adjustment 125
  whole economy time series 122–8, 132
 unconditional 28
 see also Unemployment

Demographics, unemployment and 100
Depreciation 153
Disequilibrium 49–69
 demand for labour 55–8
 empirical results 61, 62, 63, *64*, 65
 estimating models 60–1
 multi-market model 37–9, 52–4
  Belgium 181–4
 policy implications 65, 66, 67, 68
 single market model 33–5, 50
 supply of labour 58–60
Dixon, H. 155n
Dolado, J. J. *et al.* 184
Dreze, J. H. 181–4
DRI model, Data Resources Canada 175n
Driehvis, W. 184
Dunlop, J. T. 41

Earnings related supplement 114, 116
Economic Forecasting Division, Canadian Department of Finance 175
Economica 169n
EEC 170
Efficiency wage models 11–12, 41–2
Elasticity
 of substitution 26–7, 36, 44, 184n
 of supply and demand 125, 126, 147
 of unemployment 123
 real wage 29, 86
Eliasson, G. 176n
Empirical results
 disequilibrium models 61, 62, 63, *64*, *65*
 imperfect competition macromodel 160–6
 UK labour market, new classical model 121–2
Employer risk averse 12
Employment *64*
 aggregate 34
 business and government *73*
 female *21*, 22
 in imperfect competition 147–9
 male *21*, 22
 profit maximizing level 2
 real wages and 185–8
 response to shock 148, 156–7
 self *21*, 22
 simulation *90*
  government drop 95–6, *97*
 see also Participation

Endogeneity 127
Entrepreneurship 106, 120
Equilibrium, definition 49
EUROLINK model 176n
European Economic Community *see* EEC
Exchange rates 152, 153, 155n
  index, real multilateral 102
Exogeneity 154
  demand as exogenous 154
  test 127, 133
Expectations 16, 17, 30, 186
  adaptive 33
  forward looking 3
  price 86–7
  rational 33, 63
    hypothesis 55–8

F-test for equality of coefficients 126
Fair, R. C. 50, 51
Federal Reserve Board 170, 174n, 176n
Females
  average earnings 77, 78
  participation 76, 77–8
  pay increase simulation 96, 97
  supply of 85–6
Fisher, S. 15
FOCUS, Forecasting and User Simulation Model 175n
Franz, W. 176
Friedman, M. 14, 15, 18
Friedman-Phelps inflation-adjusted Phillips curve 17–18
Full information maximum likelihood 42

Gagliardi, F. 191
Geary, P. T. 42, 43, 185–6
Goldsmith, A. H. 181
Gordon, R. J. 171, 190–1
Gourieroux, C. *et al.* 53
Grady Economics and Associates Ltd 175
Greenhalgh, C. 6n
Gregory, R. G. 176, 177–8
Grossman and Hart model 12–13
Grossman, H. 37, 43
Grossman, S. J. 12
Grubb, D. 164, 172, 185n, 187
Grubb, D. *et al.* 187
Gylfason, T. 189

Hall, S. G. 49–69
Hall, S. G. *et al.* 63
Hamada, K. 176
Hansen, B. 34
Hart, O. 12, 155n
Hart, R. A. 29, 189, 190
Haugh, L. D. 185n
Hazledine 27
Helliwell, J. *et al.* 175n
Henry, S. G. 128
Henry, S. G. B. 49–69
Henry, S. G. B. *et al.* 141
Herfindahl index of concentration 147
Hicks IS/LM model 14
Hilliard, G. 124, 125
Hines, A. G. 120
HMT model *see* Treasury Model
Holden, K. 123
Holland 184

Imperfect competition model *see* Competition
Implicit contract models 12–13
Incomes policy 63
Indexation 124, 178, 191–2
Inflation
  contract behaviour and 124–5
  cost-push theory of 120
  expected 153
  monetarist unemployment and 15, 16
  Phillips curve 17–18
  unanticipated 106, 119
  uncertainty 89
  unemployment tradeoff 157–9, 162, 166
Information maximum likelihood 42
Institutional characteristics 188–93
Interest rates 58, 62
INTERLINK model 175n
International Monetary Fund 170
International perspective 169–97
  institutional characteristics 188–93
  multi-country studies 184–8
    comments 170–3
  multimarket disequilibrium model, Belgium 181–4
  standard model
    Australia 176–8
    Canada 175
    OECD 174–5
    off-model applications 176
    USA 174–5, 178–81
Intertemporal substitution hypothesis 58–9, 181

Ireland 186
Italian Wage Supplement Fund 189

Jackman, R. 165
Jaffee, D. 50, 51
Jenkinson, T. 164
Johnson, G. 150
Junankar, P. N. 123

Kennan, J. 42, 43, 185–6
Keynes, J. M. 7n, 14
Killingsworth, M. R. 31
Kniesner, T. J. 181
Knight, K. G. 7, 12n, 19n
Kochin, L. 123
Konig, K. 176
Kurosaka, Y. 176

Labour
  cyclically adjusted share 36
  marginal product *see* Marginal productivity
  rearranged competitive supply 40
  see also Demand for labour; Labour market; Supply of labour
Labour market
  concept of 1–2
  demand *see* Demand for labour
  disequilibrium *see* Disequilibrium
  identities 88, 89
  imperfect competition and 145–67
  macroeconomics and 14–17
  macromodel 153–60
  neoclassical model *see* Neoclassical model
  new classical model *see* New classical model
  opportunity cost of labour 12
  short side constraints 54–5, 64
  supply *see* Supply of labour
  tightness, degree of 34–5
  see also Models
Lags 29, 57, 58, 60, 127, 137, 185n
Lambert, J. P. 34
Lancaster, T. 135
Layard, P. R. G. 29, 39–40, 79, 178n, 184, 187
Layard, P. R. G. *et al.* 19n
Layard, R. 4n, 8, 142, 150, 160, 161, 162, 164
Layard, R. *et al.* 142
LBS *see* London Business School model
Least squares 42
Leijonhufvud, A. 49
Leisure 5, 6, 59, 78, 107
Lewis, H. G. 124
Lindbeck, A. 189
Linear dynamic simultaneous equations models 42, 195
Lippman, S. 7
Lipsey, R. G. 159
Liverpool Model 105, 110, 122, 128, 135, 136, 174n
London Business School model 174n
Loyalty payments 11
LPL *see* Liverpool Model
Lucas, R. 16n
Lucas, R. E. 27, 32, 120, 175, 178, 179, 181, 193
Lucas-Rapping model 32–3, 120, 175, 178, 179, 181, 193
Lucas-Sargent supply function 16

McCall, J. 7
Macdonald, I. M. 13, 40, 43
MACE, University of British Columbia model 175n
MacKay, D. E. 135
Macroeconomic Modelling Bureau (ESRC) 174
Macromodel 153–60
Maddala, G. S. 34, 51
Maki, D. 123
Malinvaud, E. 15, 37, 53, 176
Marginal productivity 184n
  marginal revenue product 28–9, 147, 150, 155
Market, definition 1
Market-clearing model 32–3
MCM model *see* Federal Reserve Board
Metcalf, D. 189, 191, 193
Minford, A. P. L. *et al.* 59, 65, 129, 137, 139, 142
Minford, P. 105–44, 160
Minford model 33
Mirman, L. J. *et al.* 147
Misery index 192
Mitchell, 176
Models 25–46
  bargaining *see* Bargaining
  Canadian 175n
  COMET model 176n
  CUBS 81, 88, 174n
  disequilibrium

estimating 60–1
multi-market model 37–9, 52–4, 181–4
single market model 33–5, 50
efficiency-wage models 41–2
EUROLINK model 176n
imperfect competition *see* Competition, imperfect
INTERLINK model 175n
Liverpool Model 105, 110, 122, 128, 135, 136, 174n
London Business School model 174n
Lucas-Rapping model 32–3, 120, 175, 178, 179, 181, 193
macromodel 153–60
market-clearing model 32–3
Minford model 33
MPS model *see* Federal Reserve Board
neoclassical 71–104
new classical 105–44
NIESR model 128, 141, 174n
non-structural models 42–4
right-to-manage model 40
standard model 30–2
international perspective 173–6
Treasury Model 174n
union monopoly 41
vector autoregressive model 42, 43, 44, 56, 181
wage-gap model 35–7
Modigliani, F. *et al.* 184
Monetarism 14, 15–16
Monetary shosks *see* Shocks, monetary
Money illusion 14–15
Mortensen, D. T. 32
MPS model *see* Federal Reserve Board
MTFM, Medium-Term Quarterly Forecasting and Simulation Model 175n
Muellbauer, 37
Multimarket models *see* Disequilibrium
Mulvey, C. 110, 124

Nadiri, M. I. 29
NAIRU 18–19, 191
NAPRU 18n
Nash bargain 40
National Institute of Economics and Social Research *see* NIESR model
National insurance charges 136–7
Neftci, S. N. 185, 186
Nelson, F. D. 51
Neoclassical model 71–104
demand for labour
by business sector 81, 82, 83, 84
theory 72–5
econometric results 81–9
simulations *see* Simulations
supply of labour
female 85–6
male 84–5
theory 75–9
wage behaviour 79–80
New classical model 105–44
closing open economy model 108–10
criticisms of 139–42
demand for labour *see* Demand for labour
labour market structure 110–1
outline of 106–10
supply of labour, theory 111–20
New Zealand 186
Newell, A. 172, 173, 187, 191, 192,
Nickell, S. 59, 65, 75, 121, 122, 135, 140, 142, 145–67
Nickell, S. J. 29, 34, 39–40, 41n, 79, 174, 178n, 183n, 184, 187
Nickell-Andrews equations 81
NIESR model 128, 141, 174n
Non-accelerating inflation rate of unemployment *see* NAIRU
Non-accelerating prices rate of inflation *see* NAPRU
Non-structural models 42–4
North Sea oil 155, 162

OECD 169n, 170
standard model 173–4
unemployment 170–1
Oil
North Sea 155, 162
price shocks 27, 178
Okun's law 105, 126
One-step-ahead simulations *see* Simulations
Open economy model *see* New classical model
Organisation for Economic Cooperation and Development *see* OECD
Oswald, A. J. 7, 13, 117
Output
changing expected level of 67
planned 121
value added 45, 46, 145–6

Parsley, C. J. 124
Participation 102, 112
  decision 5, 6, 31, 32
  female 77–8
  ratios, male and female *76*
Peel, D. A. 123
Pencavel, J. H. 123n
Phelps, E. S. 15, 32
Phillips curve 17–18, 40, 71, 128, 172, 174
  augmented 79, 177
  concavity *159*, 160
  expectations-augmented 30, 186
  in disequilibrium models 128
  inflation-adjusted 7–18
  market-clearing concept 32
  OECD 185n
Pichelmann, K. 176
Portes, 37
Poverty base lines 127
Price, R. 116
Prices
  average cost 147
  countercyclical mark-up 147, 149
  expectations 86–7
  factor price ratio *74*, *75*
  in disequilibrium approach 50
  in imperfect competition 145–7, 148–9
  inflation and 157–8
  unresponsive behaviour 53
  value-added 45, 46, 145–6
Production function 55, 56
  as supply of output equation 106
  CES 26–7, 36, 44, 184n
  Cobb-Douglas 26–7, 28, 36, 44, 175, 183
Productivity 11, 140
  marginal *see* Marginal productivity
Purdy, D. L. 120

QFS, Quarterly Forecasting and Simulation Model 175n
Quandt, R. E. 33, 34, 52, 53, 55, 178

Random walk hypothesis 180
Rapping, L. A. 27, 32, 120, 175, 178, 179, 181, 193
Rational expectations *see* Expectations, rational
RDX2, Research Department Experimental Model 175n
RDXF, Research Department Experimental Forecasting Model 175n
Reid, G. L. 135
Replacement ratios 5–6, 17n, 59, 112, 113n, 123, 135, 162
Retention ratio 19, *20*
Retirement benefits 76, 84
Revenue function *see* Tax
Right-to-manage model 40
Risk aversion, employer 12
Rosen, H. S. 33, 34, 55, 178
Rosen, S. 29
Rossi, V. 176n
Rotemberg, J. J. 147, 187, 190

Sachs, J. 19, 73, 190, 191–2
Sachs, J. D. 28, 35, 36, 176, 178, 186, 187
Sales constraint 55
Saloner, G. 147
SAM, Small Annual Model 175n
Santomero, A. M. 31
Sargan, J. D. 128
Sargan-type equation 128, 134
Sargent, T. J. 16n, 30, 33, 57, 68, 105, 117, 185
Search models 8–11, 111
  employer search 10
  worker search 8–10
Seater, J. J. 31
Shapiro, C. 11n
Sherriff, T. D. 120
Shocks 148, 156–7, 186
  monetary 15, 16
  oil price 27, 178
  stagflation and 191
Sims, C. A. 42, 185n
Simulations, 89–100
  dynamic
    benefits increase *95*, *96*
    capital stock increase 93, *94*
    female pay up *96*, *97*
    governmental employment drop 95–6, *97*
    income tax cut *93*
    structural change 94, *95*
    unemployment causes *98*, 99, 100
  one-step-ahead 89–93
    earnings growth *91*
    employment, static *90*
    errors 89, 90

unemployment rate, static *91*
vector autocorrelation matrix 90, 92, 93
Single market disequilibrium model 33–5, 50
Smith, R. 127
Smith/Wu test 127
Sneesens, H. 37–8
Sneesens, H. R. 181–4
Solow, R. M. 13, 15, 40, 43, 171, 173, 193
Soskice, D. 155n
Spillover 38, 52–3, 54
Spindler, A. 123
Standard model 30–2
international perspective 173–6
Stern, N. H. 11n
Stiglitz, J. E. 7, 11n, 150
Structural change 84, 85, 102
simulated *94*, *95*
Summers, L. 150, 164
Summers, L. W. 19
Supply of labour 3–7, *64*
disequilibrium approach 58–60
dynamics of equation *59*, 60
intertemporal substitution hypothesis 58–9, 181
married women 6–7
neoclassical model
female 85–6
male 84–5
theory 75–9
new classical model 111–20
elasticity 125, 126
unemployment benefits and 4–6
Supply-side economy 107, *108*
Symons, J. S. V. 83, 172, 173, 178, 187, 191, 192

Tax 2, 100, 125, 135, 136, *137*, 151, 162, 187
allowances 103
NTAX 63
simulated cut *93*
Taylor, J. B. 43, 187
TIM Informetrica Model 175n
Time series 122–8, 132, 191
Townsend-Greenspan Inc. 174n
Trade *see* World trade
Trade unions
bargaining 189
models 12–13
union power 58, 59, 62, 63, 105, *141*, 189
unionization 127, 135, 136, *137*, *141*
new classical model and 106, 110–11, 118, 123
wages 112, 117, 118, 151
Transversality condition 117
Treasury Model 174n
Treble, J. G. 123, 124
Trivedi, P. K. 178
Tzannatos, Z. 72, 96

Unemployment *21*
benefit *see* Unemployment benefits
duration 112, 135, 165–6
equations 122, 130, 131
female *20*, 22
hysteresis model 19
in wage/price model *158*
inflation tradeoff 157–9, 162, 166
inflows and outflows from register 126
Keynesian 14–15
male *20*, 22
models 7–13
efficiency wage 11–12
implicit contract 12–13
search 8–11
trade union 13
monetarist 14, 15–16
NAIRU 18–19, 191
natural rate 18–19, 71, 155n, 158, 162, 163
new classical model 135–6, *137*, 138–9
outsiders 13
reserve army 11
simulation *91*
UK (1972–85) *98*, 99, 100
voluntary 7–8, 33
see also Demand for labour
Unemployment benefits 58, 59, 62, 100, 135, 136, *138*
in new classical model *115*
labour supply and 4–6
real 19, *20*
reduction effects 66, 67
Union power: unionization *see* Trade unions
Union-monopoly model 41
United States 171, 185, 188
macroeconomic models 174–5
non-structural evidence 178–81

Utility functions 4, 5, 8

Vacancies *21*, 22
Value-added output 45, 46, 145–6
Varian, H. R. 4n
Vector autocorrelation matrix 90, 92, 93
Vector autoregressive model 42, 43, 44, 56, 181

Wage equation 134
  accelerationist 31
  industry 151–2
  real 122, 130, 131
  reduced-form competitive 40
Wages
  adjustment 34, 63, *65*, 86, 87, 88–9
  Australian model 177
  average or hours adjusted earnings 83
  bargaining *see* Bargaining
  consumption 30, 32, 58, 151
  determination in imperfect competition 149–52
  efficiency wage model 11–12
  female
    average earnings 77, 78
    simulated increase 96, 97
  free market 112
  gap 35–7, 179
  indexation 124, 178, 191–2
  insider 150, 164
  neoclassical model 79–80
  nominal, cost-push theory of 120
  normal real 33, 120
  offers 8–9
  outsider 150–1
  Phillips curve and *see* Phillips curve
  pre-tax real 30
  pressure variables 157
  product 30, 36, 151
  real 19, *20*, 30
    and employment hypothesis 185–8
    consumption wage 30, 32, 58
    elasticity 29, 86
    expected normal 33, 120
    product wage 30, 36
    sluggishness 37
  reservation 9, 10
  response to shock 148, 156–7
  rigidity 31, 43, 116, 186–7
  simulation *91*
  stickiness 14, 15, 19
  union 112, 117, 118, 151
Wagner, M. 176
Wallace, N. 16n
Wallis, K.F. *et al.* 44, 174
Warburton, P. 3, 19n, 71–104
Waterson, M. 147n
Weiss, A. 11
Wharton Econometric Forecasting Associates 174n
Whitbread, C. 100, 101
Whitley, J. 62
Williams, C. 165
Working population 58, 62
World Bank 170
World trade 123, 136, *139*
  terms of trade 162
  trade balance 154–5, 162
  trade cycle 17
Wren-Lewis, S. 155n
Wu-test of exogeneity 127

Zabalza, A. 72
Zis, G. 120

www.ingramcontent.com/pod-product-compliance
Ingram Content Group UK Ltd.
Pitfield, Milton Keynes, MK11 3LW, UK
UKHW020628120926
15112UKWH00034B/115
* 9 7 8 9 4 0 0 9 1 2 0 4 5 *